Praise for *A Field Mc*

We successfully transformed the C(
prototypes of the methods described in this field manual. Read it, follow it, and hold on for what can be both a rapid and effective approach to deep student (and faculty) engagement and success.

—**Gary Bertoline**, Senior Vice President for Purdue Online & Learning Innovation

Dipping and diving into the methods described in this field manual has helped me to unleash engineering students towards building a sociotechnical leadership mindset. The methods reach beyond course content and provide practical insights that can start us down the path to making the changes that higher education needs today.

—**John R. Donald**, Past President, Canadian Engineering Education Association
and Director, University of Guelph Engineering Leadership Program

How did the MIDAS programme at UTS do it? By adopting the heart and mind approach advocated in A Field Manual for a Whole New Education, *our faculty and students embraced a once voluntary studio-based approach with energy. Read and re-read this practical guide and achieve startlingly effective changes today.*

—**Beata Francis**, 'Heart & Mind,' UTS MIDAS

I became aware of the possibilities unleashed by this field manual through a pioneering initiative to spread its methods around the globe. As an engineer at a major consulting firm in Latin America, I had the opportunity to join a pilot where I had my practice and my life so deeply changed that it led me to direct a successful initiative to help poor-smart-altruist undergrads learn how to lead. This manual is a must read for everyone who wants to lead the educational change that the world is urging!

—**Germano Garrido**, Chairman and Former Executive Director, Instituo Semear

This field manual is chock full of innovative thinking and specific, practical action frameworks. Read it, reflect upon it, and use it to reboot your corner of higher education culture.

—**Beverly Jones**, Author of *Find Your Happy at Work*

A Field Manual for a Whole New Education *builds on the groundbreaking work of* A Whole New Engineer, *Goldberg and Somerville's previous book. It highlights the importance of what they call 'shift skills' including reflection, curiosity, and listening. The "reflexercises" provide a practical method for integrating the learning in this important contribution to the ongoing "reboot" of higher education.*

—**Pat Mathews**, Past Director, Georgetown Leadership Coaching Program

Witty and briskly written, this book is a whirlwind of insights into why conventional academic processes of educational reform often end up just promoting curricular tweaks rather than motivating creative, inspirational change. Building on the success of their A Whole New Engineer, *the authors bypass conventional memes to develop an ambitious process involving activities such as curious listening, uplifting emotion and joy, and taking student perspectives seriously. A highlight of the volume is the many opportunities it presents for readers to engage with it at their own pace and level of interest. An ambitious and refreshing must-read not only for engineers but for anyone interested in practical how-to's for transforming programs in higher education.*
—**Diane P. Michelfelder**, Professor of Philosophy, Macalester University

A Whole New Engineer *was my everyday companion in bringing effective change to a well-regarded Brazilian engineering school. After looking at this new text, I am confident that* A Field Manual for a Whole New Education *will be my everyday companion as I work to bring needed changes to a plural and venerable Brazilian public university. Said more simply and in Portuguese,* o livro é muito legal!
—**Alessandro F. Moreira**, Vice-Rector, Universidade Federal de Minas Gerais

Educational systems are notoriously resistant to change. But this smart and timely book offers practical advice for overcoming inertia and plowing a path to progress. Read it and help improve education today.
—**Daniel H. Pink**, #1 New York Times bestselling author of *The Power of Regret, Drive,* and *A Whole New Mind*

Critical aspects of A Field Manual for a Whole New Education *are proving very supportive for our transformation journey in DCU Futures, where our ambition is to radically reimagine the undergraduate curriculum. It prompts us to reflect, challenges our orthodoxies, and provides structures and frameworks that empower us. In short, it is a go-to text that is never far from my reach, and I unequivocally recommend it.*
—**Blánaid White**, Dean of Strategic Learning Innovation, Dublin City University

Praise for *A Whole New Engineer*

A Whole New Engineer *provides excellent ideas about how to make engineering colleges more motivationally aware and supportive of their students' development.*

—**Edward L. Deci**, University of Rochester, author of *Why We Do What We Do*

Much debate about educational reform is cast in strictly rational terms. It's both ironic and moving that a couple of engineers have nailed the importance of love, empathy, and caring as foundational in rethinking education.

—**Mark Goulston**, author of *Just Listen* and *Real Influence*

As people change careers or occupations later in life, it takes spirit, strength, and a thirst to find meaning in this life. A Whole New Engineer *explores a new kind of education where the courage to do so is cultivated earlier and with greater purpose.*

—**Kerry Hannon**, author of *What's Next?*

Everyone involved with educating engineers should read this book twice—once for inspiration and a second time for planning. A Whole New Engineer *provides a road map for overhauling the stale, soul-deadening, plug-into-the-equation style of engineering education. In its place will emerge a system built on creativity and collaboration and initiative.*

—**Dan Heath**, co-author of *Made to Stick, Switch,* and *Decisive*

This isn't just a book about engineering. It's a book about education, entrepreneurship, and—ultimately—the future. Read it and prepare to take notes!"

— **Daniel H. Pink**, author of *To Sell is Human, Drive,* and *A Whole New Mind*

Much organizational change is blind to culture and emotion, and as a result, much change is ineffectual. A Whole New Engineer *approaches the change of engineering education with new cultural assumptions based on the power of experiential education and the involvement of the learner in the process.*

—**Edgar H. Schein**, MIT Sloan School, author of *Humble Inquiry*

Far more than an inspiring story about the transformation of engineering education, this remarkable book serves as a call to action and blueprint for reinventing all education for the twenty-first century. Everyone concerned about the preparation of the next generation for their—and our—future should read this important book.

—**Tony Wagner**, Harvard University, author of *Creating Innovators*

A Field Manual for
A Whole
New Education

A FIELD MANUAL FOR

A WHOLE
NEW EDUCATION

Rebooting Higher Education for
Human Connection and Insight
in a Digital World

DAVID E. GOLDBERG
with MARK SOMERVILLE

ThreeJoy Associates, Inc. | Douglas, Michigan

Published by
ThreeJoy Associates, Inc. | Douglas, Michigan

Publisher's Cataloging-in-Publication Data
Goldberg, David E. (David Edward), 1953-

A field manual for a whole new education : rebooting higher education for human connection and insight in a digital world / David E. Goldberg with Mark Somerville. – Douglas, MI : Threejoy Associates, Inc., 2023.

p. ; cm.

ISBN13: 978-0-9860800-5-0

1. Engineering--Study and teaching (Higher)--Handbooks, manuals, etc. 2. Technical education--United States. 3. Engineering schools--United States. 4. Educational innovations. 5. School improvement programs. I. Title. II. Somerville, Mark, 1967-

T73.G65 2023
620.001--dc23

Project coordination by Jenkins Group, Inc. | www.jenkinsgroupinc.com

Cover design by Yvonne Fetig Roehler

Interior design by Brooke Camfield

Printed in the United States of America

27   26   25   24   23   •   5   4   3   2   1

For the beautiful opportunity to
bring wholehearted change to higher education

Contents

Figures

Tables

Preface
A Few Words about This Field Manual

A Word about the Origins of the Project, Authorship, and Voice

For some number of years, Dave and Mark had conversations about writing a practical workbook to follow up on the earlier volume, *A Whole New Engineer: The Coming Revolution in Engineering Education* (Goldberg & Somerville, 2014). Dave started and stopped writing such a workbook several times between 2014 and 2020, but he struggled each time to (1) carry over the lessons of the earlier volume while (2) generalizing the result to all higher education. Then in 2020, while writing a new introduction to the Chinese translation of *A Whole New Engineer* (available from UESTC Press, *http://wholenewengineer.org/chinese-4/*), it became clear that the topics in that new introduction made an excellent outline for the organization of this field manual.

At that crucial moment of insight and clarity, unfortunately for the project (and fortunately for Olin College), Mark Somerville accepted an interim (and subsequently permanent) role as Olin's provost; as such he was unavailable to play an active role in the writing of the field manual, but he did agree to support the project by offering Dave counsel and advice at critical junctures during the project's evolution. At that time, both Dave and Mark agreed that that continuity between the two projects—*A Whole New Engineer* and *A Field Manual for a Whole New Education*—demanded sticking to the collaborative "we" voice of the first volume, and henceforth the field manual uses that voice throughout.

Similarities and Differences Between the Two Projects

In the preface to *A Whole New Engineer*, we wrote the following:

> *As academics, we are accustomed to writing for an audience of our peers—others who speak the language of academic research.* A Whole New Engineer *is a different animal because we are addressing not only fellow academics but also a wide world of others—students, parents, educators, employers, practitioners.*

We went on to say that that book was not a research text, but rather a memoir of our experiences and conversations around change in the education of engineers, a narrative driven by little models of transformation as well as by stories—historical stories, stories of failure, stories of success, stories of unleashed students and faculty. Our hope was to motivate readers to think about how engineering education might be different, to have challenging, perhaps inspiring conversations that might lead to engineering education better aligned with our times.

Judging by book sales and many of the conversations we've had and heard, our hopes have largely been realized, and we've been wondering over the intervening eight years how we might further aid those toiling in the vineyards of higher-education change. These reflections have led to the aims, methods, and the writing of this field manual.

First, we largely abandon the pervasive storytelling style of *A Whole New Engineer* and become more how-to and practical. This is after all *A Field Manual for a Whole New Education* and we intend for it to be used in day-to-day practice by those working to make change in the field.

Second, we enlarge the scope of the "field" we address by aiming at all higher education. We will say more about this in the opening chapter, but here suffice it to say that we believe the methods of this book can—perhaps should—be used beyond engineering in higher education more generally. Some have encouraged us to enlarge the scope of the manual even further to all education, and we suspect that the techniques of the field manual may be used in primary and secondary education, but neither of us is experienced enough to tackle those earlier educational experiences with confidence.

Acknowledgments

A book passes through many hands before it is published, and this field manual is no exception. We are grateful to the team at the Jenkins Group including Jerry Jenkins, Jim Kalajian, Leah Nicholson, and Yvonne Roehler for their professional production work in bringing the manuscript to life.

We are grateful to those who took the time to review the manuscript and offer testimonials. We thank Gary Bertoline, John R. Donald, Beata Francis, Germano Garrido, Beverly Jones, Pat Mathews, Alessandro Moreira, Daniel H. Pink, and Blánaid White.

Dave thanks his ThreeJoy coaching associate Emma Schoenfelner for her hard and smart work in facilitating the Big Beacon radio podcast, for producing the ThreeJoy Coaching Club, and for keeping the various websites and digital platforms of threejoy.com and bigbeacon.org up and running. These platforms and activities led by twists and turns to much of the content of the field manual. Dave thanks Joe Murphy for important legal advice during the writing of the book.

Work with many coaching clients—including faculty members, academic leaders, and practitioners—has shaped how Dave thinks about personal and organizational insight and change, and his coaching clients are protected from individual acknowledgment by the veil of coaching confidentiality. Nonetheless, you know who you are, and one of the great joys of Dave's life has been to be trusted by you with the intimacy of the deeply meaningful conversations shared by client and coach. Likewise, Dave thanks his coach, Bev Jones, for her penetrating questions and on-target homework assignments. As former Google chairman, Eric Schmidt said, "Everyone needs a coach," especially coaches themselves, and Dave is grateful for Bev's grace, skill, wisdom, and insight.

Extended change initiatives at a variety of institutions were the hot houses where many of these methods were piloted and refined. Thanks are due to the leaders, faculty members, and student participants in a variety of initiatives. These include (in rough order of engagement) University of Illinois, National University of Singapore (NUS), Tembusu College (NUS), Purdue Polytechnic Institute (Purdue University), Saugatuck Center for the Arts, University of Technology Sydney, Universidade Federal de Minas Gerais (UFMG), Falconi Consultants, Universidad de los Andes (Uniandes), and Dublin City University (DCU). Many of the modules used in these initiatives grew out of the TASL (Teaching as Servant Leadership) and POCA (Personal and Organizational Change) workshops Dave piloted at NUS for the Design Centric Curriculum in 2011 and 2012, and those modules were further refined in the *Change that Sticks* summer workshops run for several years at Olin College by Dave and Mark.

Opportunities to give workshops and talks, participate on panels, and have interactions at institutions and conferences have advanced Dave's understanding of higher education change and how to communicate it, both. Dave acknowledges a number of those experiences (in alphabetical order): Afeka College, African Engineering Education Association International Conference (Lagos), Chinese Higher Education in Engineering Summit (Chengdu), Delft University of Technology, International Exhibition & Conference on Higher Education (Riyadh), James Madison University, the Keen Network, Lassonde School of Engineering (York University), LASPAU (Harvard), McMaster University, Memorial University of Newfoundland, Ngee Ann Polytechnic, Olin College, Politecnico di Milano, Pontificia Universidad Católica de Chile (UC), Pontificia Universidad Católica del Perú (PUCP), Proctor & Gamble, Ramon Llull University, Republic Polytechnic, Singapore Polytechnic, Texas A&M University at Qatar, Universidad de Ingeniería y Tecnología (UTEC), Universidad de Piura, Universidad Técnica Federico Santa Maria, University of Electronic Science and Technology of China, University of Glasgow Singapore, University of

Twente, Waseda University, University of Cincinnati, University of Lagos, and the World Engineering Education Forum (Chennai).

Dave thanks his collaborators John Donald, Bea Francis, and Germano Gerrido for their support, interaction, and insights over the years. Dave thanks his philosophy colleagues Diane Michelfelder and Ibo van der Poel for continued interactions. He is grateful to new fPET (Forum on Philosophy, Engineering & Technology) acquaintance Nina Jouraskova for helping Dave get his writing mojo back during a recent collaboration on a surprisingly tricky paper about coaching, engineering, and philosophy. Dave thanks Kristin Armstrong for the chance to try many of these methods in a local, non-profit organizational setting. Dave thanks Gary Bertoline and Alessandro Moreira for ongoing and important conversations about educational change over the last decade or so.

Penultimately, Dave wants to acknowledge that meeting Mark Somerville in 2008 was one of the great blessings of his life. Since that time, Dave has counted on Mark as a steady and wise voice and partner in the evolution of iFoundry, the Olin-Illinois Partnership, Big Beacon, *A Whole New Engineer,* and the *Change that Sticks* workshops. Dave thanks Mark for not giving up on this project even though he was unable to jump right in. Nonetheless, many times during the writing Dave asked himself WWMD ("What would Mark do?") and came to good answers in the imagined conversation; and when that wasn't good enough, Mark was there for interactive reflection, great framing questions, and pointed, yet kindly, guidance.

Finally, Dave met Mary Ann Goldberg (nee Weir) 45 years ago on his one and only blind date. Dave thanks Hank Aberman for the set up, and he expresses his gratitude and love to his wife and life partner of 40+ years. She was unwavering in ways that were and are an inspiration.

Mark acknowledges and thanks the entire Olin community: faculty, staff, students, parents, and friends of Olin, past and present. Being a part of Olin—from the team that helped Olin become reality to the community Olin is today—has been transformative for him, and many of the lessons distilled here have roots in those experiences.

Mark also wishes to acknowledge and thank the many different schools and organizations he has collaborated with over the years. In addition to the obvious thanks to the College of Engineering at the University of Illinois, he also particularly wants to thank Delft University, the University of Texas at El Paso, the Woodrow Wilson Academy, the Kern Family Foundation and the KEEN Network, the Argosy Foundation, and Fulbright University Vietnam. The opportunity to be part of these organizations' journeys was incredibly rewarding and enlightening.

There have been many teachers, friends, and colleagues over the years who asked the right questions or listened without judgment during challenging times. A very incomplete list includes Melvin Oakes, James Duban, and Christine Maziar at UT Austin; Jesús del Alamo and Terry Orlando at MIT; the entire Department of Physics and Astronomy at Vassar College; the early members of Active Learning in Engineering Education; friends from other institutions including Aldert Kamp, Ben Wilkinson, Sebastian Dziallas, Ryan Derby-Talbot, Megan Bulloch, Dinh Vu Trang Ngan, Roger Gonzalez, Shane Walker, Meagan Vaughn, and Deborah Hirsch; and of course many friends from Olin including John Geddes, Jon Stolk, Zach First, Caitrin Lynch, Ela Ben-Ur, Jeff Goldenson, Aaron Hoover, Lawrence Neeley, Linda Canavan, Susan Mihailidis, Joanne Pratt, Alisha Sarang-Sieminski, Michael Moody, Benjamin Linder, Vin Manno, Gill Pratt, Lynn Stein, Jessica Townsend, Sam Michalka, Jason Woodard, Allen Downey, Stephen Schiffman, Sarah Spence-Adams, and Rob Martello.

Finally, Mark thanks his wife Babette and daughters Charlotte and Josephine. Their support and love made all the difference.

1

To Higher Education with Love

This text is intended as a pithy field manual for the recrafting of higher education to align with the imperatives of our fast-paced age. Challenges to the future of work abound in our times, an era teeming with increasingly competent machine learning (ML) and artificial intelligence (AI), with ever more mobile and functional robots, with myriad apps disrupting one industry after another. Outlining *what* and *how* higher ed should change in response to this *digital gaggle* is a daunting task; given that engineering and technology are responsible for much of the upheaval, it is interesting and a bit ironic that we turn to the education of engineers for guidance on how to return human connection and insight to higher education to stave off the onslaught of the technology wrought by engineering hands.

In particular, this book builds on the learning and approach presented in *A Whole New Engineer: The Coming Revolution in Engineering Education* or WNE (Goldberg & Somerville, 2014). WNE was published in 2014 as both a change memoir and travel guide to transformative change in engineering education. The present volume (*A Field Manual for a Whole New Education* or simply *the field manual*) updates our learning journey and carries the learning of WNE over to higher education more generally by outlining a practical approach of (1) what to change to both fend off and

1

co-exist with the digital gaggle and (2) how to change significantly, quickly, and with wholehearted embrace.

Transforming Engineering Education with Emotion & Culture

The manual starts by revisiting some of the key concepts of WNE and what we've learned since its publication. WNE was published in late 2014 and immediately started to sell well in engineering education circles around the world. Early on we started hearing stories of deans and department heads of engineering walking around various campuses displaying their copy of the book, telling their colleagues and faculty members things such as, "We need to do more of this here!" We heard stories of independent-minded faculty members and students using the book to hold reading clubs and groups from Doha to Singapore, from Enschede to Belo Horizonte, and back to West Lafayette, Toronto, and Tucson. This is exactly the kind of impact we had hoped for, but why has the book been landing as it has? What was it that got people talking and acting in useful, new ways? We had a sense of the mechanism of possible acceptance and adoption when we wrote the following in the original introduction to the book [WNE, p. xxiii]:

> *As we reflected on these experiences* [at Olin and Illinois], *we came to recognize that our initial thinking about the keys to educational reform was wrong. The key variables weren't pedagogical. They weren't financial. They weren't curricular. They weren't research. They weren't any of the usual things we've always talked about as the engines of change. The variables were deeply* <u>emotional</u> *and* <u>cultural</u>.

In looking back, one of the reasons the book has generated the excitement it has, because it has breathed fresh air into the global engineering education change conversation by suggesting that maybe engineering education has been looking for change in all the wrong places. By focusing primarily on content, curriculum, and pedagogy, perhaps engineering education has been missing the emotional and cultural boat that could make an engineering education more holistic, more entrepreneurial, more agile, and more authentically fulfilling for more different kinds of people. In a sense, looking back, the book has given—and is still giving—people permission to have different conversations around the nature of the question "What is a good engineering education?" It invites deeper, more open-ended conversations that create a new possibility space for experiments or little bets that can lead to effective action and scale up.

Does This Work for Higher Education More Generally?

All this is well and good for engineering education and engineering educators, but what has any of this to do with other flavors of higher education? The answer, we believe, is that the lessons learned in engineering education with engineering educators and students transfer well to higher education, educators, and students, generally. For one thing, much of what we discuss in the book is process oriented, and thus is as applicable to engineering as it is to animal husbandry or English literature. But as we observed above, a key insight of *A Whole New Engineer* was to shift and then balance emphasis on content, curriculum, and pedagogy with increased concern for emotion and culture, and in this regard, we may view engineering education as something of a bounding case study. If you can successfully help individuals and organizations steeped in the rational traditions and rigorous content of an engineering education shift and balance toward emotion and culture, you can help others find their way back to a more human education, even in a time increasingly overrun with the artifacts of and consumed with attention to the digital gaggle.

Another point to make involves *vocation* and *practice* in the world. One of the charges routinely leveled at engineering education is that it is too oriented toward the profession and vocation of engineering, but it is odd, and a bit concerning, that the charge is less frequently leveled at liberal education that it is too little concerned with vocation and practice in the world. This is not the place to consider the depth and breadth of a modern engineering degree. Nor is it the place to fully explore the rigor or employment prospects of a liberal arts degree. It is interesting that increasingly the humanities and engineering in the academy make the same mistake regarding the notion of *practice*. Whereas the humanities once celebrated its understanding of the central importance of *conversation* and *reflection* about its objects of study as the critical essence of a liberal education, increasingly humanists (Toulmin, 2001) have joined engineers by viewing the teaching and the application of particular theories as the royal road to practice. We shall say more about this in chapter 3 when we talk about what we call *theory privilege* and when we develop the *five shifts* in chapter 4.

Roadmap to the Remainder

Chapter 2 summarizes the key lessons of *A Whole New Engineer* to familiarize those who have not read the book, remind those who have, and to help create a solid foundation for the new work reported herein. As mentioned above, WNE advocates cultural and emotional change to support student unleashing and autonomy, and chapter 2 succinctly summarizes key models and findings of the original text.

Chapter 3 summarizes new practical findings since WNE was published, findings that have resulted from our continued work to bring the WNE model to fuller fruition. The headline items include the importance of the rise of the digital gaggle, new earning-learning business models, rigorous soft skills (what we call *shift skills*), co-contrary thinking, the problem of *theory privilege*, accomplishing change in months not years, and the crucial role of affect in the implementation of change plans.

Chapter 4 takes the notion of shift skills from chapter 3 and defines what we call the five core shifts or simply *the five shifts*. Shifts are a combination of mindset and skill change that result in an educational environment more amenable to student trust, courage, and unleashing. The five shifts emphasize currently underappreciated skills and mindsets including reflection- and conversation-based practice, the importance of body sensations and emotional feelings, speech acts, effectuation or little bets, and co-contraries. Although our original intent was to help higher educators to coach students and facilitate better educational environments, a deeper examination of the shifts shows that they are exactly the skills that humans need in the face of the rising digital gaggle. In other words, the core shifts are the very things that computers don't do very well. In an era of rising AI and ML, going where computers don't makes sense, and here we advocate more higher educational emphasis on the five shifts. This deeper understanding of the five shifts and thinking about the interrelationship between the core shifts leads to a particular word: *insight*. Seen in this way, the five shifts are closely tied to the special ability humans have, to learn deeply through insight, something that is explored toward the end of chapter 4.

Chapter 5 elaborates on the key skills underlying reflection- and conversation-based practice by focusing on curious listening as practiced via NLQ (noticing, listening, and questioning). These are essential communication skills as taught to and practiced by executive coaches that we find extraordinarily helpful toward teaching students and faculty the underpinnings of reflective and conversational practice in school and in personal and professional life. In particular, coaches use what we call *curious listening* to help others have insight, and in this way, coaches learn about the nature of insight incubation from and for their clients, and thereafter for and from themselves. If curious listening is taught more broadly, we believe it can become the keystone habit of more widespread human insight making.

Chapter 6 builds on the five shifts and curious listening to articulate a field-tested process for *rapid* educational change with widespread and enthusiastic *embrace* by a broad cross section of stakeholders. The process— the *four sprints and spirits method* (4SSM)—draws on important practical frameworks including agile software development (Hills & Johnson, 2012), design thinking (Innovation Training, n.d.), and the five shifts to address what we call the *eight failures of traditional curriculum reform*. The method helps faculty, students, and other change stakeholders reach beyond their current biases to both preserve the best of the status quo and to innovate in ways that are likely to succeed within the culture of the educational program.

Chapter 7 takes the two outputs of the 4SSM process, (1) the educational plan and (2) the good mood and nascent culture of the change process and helps implement them both into more permanent practice. This post-4SSM process is called *affective implementation* because it pays attention to affect and the plan, both. To help preserve the mood and culture of the 4SSM process, affective implementation selects implementation team members carefully and creates a virtual or physical space, the incubator or respectful structured space for innovation (RSSI). In this way, the RSSI acts as a place to nurture both the intangibles of mood and culture as well as more tangible elements of the plan.

Chapter 8 brings the book to a close by stepping back and taking stock of its key contributions: the five shifts, the four sprints & spirits, and affective implementation. The five shifts give individuals habits of thought, language, and action to facilitate a culture of curious listening. The four sprints and spirits lead a group of stakeholders on a journey to a great educational plan and a good mood and culture; affective implementation takes both the plan and the mood seriously to bring the plan to fruition in a way that nurtures the new mood-culture and respects student-faculty workload.

Finally, the book asks readers to take the manual to the field by one of three *takes*: (1) following the sequence of the field manual in the order presented, (2) using an approach that starts with an incubator or RSSI and five shifts training, then making curricular change as it becomes feasible using four sprints and spirits in a more leisurely manner, and (3) bracketing curriculum change and using the methods of the book to work on humanizing student experience and unit culture and mood more directly. The field manual sequence (take 1) gets change most quickly, but requires careful facilitation, sequencing, modest resistance, and strong unit leadership. Take 2—the RSSI-and-five-shifts-first approach—is more tolerant of uneven process, greater stakeholder resistance, and less attentive unit leadership. Going straight at designing unleashing educational experiences by focusing on culture and mood directly (take 3) eliminates the preponderance of faculty ego involvement and works directly on the central variables that affect student engagement and learning.

Four Keys to Getting More Out of the Field Manual

Before continuing, there are four keys to getting more from the field manual, and we introduce them in contrary form (signified by italics and the double ampersand symbol &&):

1. *Dipping && diving*
2. *Knowing && not knowing*
3. *Mindset && skill*
4. *Reflection && action*

One of the curses of a university education is the tendency to learn to view reading as a means to working some problem, passing an exam, or the writing of a paper. As a result, for many professionals, reading for slow, mastery at depth—what we call *diving*—is the default. Others have learned through on-the-job experience that quick, skimming of the big picture—what we call *dipping*—may be a useful way to read these materials. Here we recommend them both, *dipping && diving* as appropriate. A quick dip of a chapter can be followed by a dive into those processes, shifts, and skills that seem to offer the most promise for augmenting your current bag of tricks. Moreover, this process of dipping and diving can be repeated in subsequent readings of the text.

A second key to using this book may be found in the contraries of *knowing && not knowing*. For example, in a later chapter, we will call out the importance of curious listening to a whole new education and some reading these words will yawn and say "Listening. Yeah, I'm a great listener." And in a certain sense those persons would be right. We are all listeners from an early age, and therein lies the value of approaching the familiar words and skills of this book with awareness around both (1) what you *know* about them and (2) what you *don't know* about them. Are there ways for all of us to go deeper as listeners? Yes, of course! Moreover, almost every concept in this book can be treated as something both familiar and unfamiliar. To approach these subjects with beginner's mind and a combination of "I know" and "I don't know" accelerates the learning process.

Much of what we discuss in this book can be treated as a matter of building skills. Returning to the earlier example of curious listening, we will emphasize listening in new ways as fundamental, and practicing those skills in educational environments is the start of a different educational practice, but skills are only a part of the equation. As we practice the skills, what we find is that there is a concomitant *mindset* shift that emerges that is as important as the skills we use. In times past, the sage-on-the-stage professor was a research and output machine; the listening professor is attuned to what his or her students are learning and when they can be given greater autonomy and charge of their learning. For some, this can be a major shift in educational mindset, but notice that it doesn't come from having an explicit value of being "student-oriented" or "student-centered." It comes from listening to students in a deeper way, and the practice leads to a mindset shift that becomes habitual.

Finally, this book will be most useful with consistent *reflection* followed by the taking of small *actions* to exercise the idea, skill, or process being considered. Peter Sims (2011), building on Saras Sarasvathy's work (2008), calls these little actions *little bets*; very small, low cost, low action trials can be a great way to bring thoughts into the world. To encourage an appropriate amount of reflection, each chapter concludes with a *reflectionaire*, a series of open-ended questions designed to get you to relate your experiences to the ideas, skills, and processes discussed in the book. Chapters also end with a list of suggested *little bets* the reader can attempt to the extent it makes sense in their change initiative. Additionally, the book has several in-line *reflexercises* that combine reflection and action in ways that relate directly to the chapter passages at hand.

In this spirit, we conclude the first chapter with our first reflexercise, the suggestion to start a new or use an existing journaling practice in the reading, reflection, and action steps that come from this book. Indeed, the book is a great excuse to exercise or start, resurrect, or modify a journaling practice so that your reflections (and results from little bets) are all in one place. However you choose to do it, the one-two punch of reflection followed by little action steps is a great way to engage the field manual.

Reflexercise 1.1: Acquire a notebook (lined or unlined, physical or virtual) to be used as a *change journal*. One way to distinguish between a diary and a journal is that a diary emphasizes what you've experienced, and a journal emphasizes your thoughts, feelings, and body sensations about what you've experienced.

How to start. Skim back over this chapter. What parts resonated with you? What parts did you disagree or have some concern with? What terms were use in ways that seemed familiar or unfamiliar? What experiences in your life connected to what you read? What questions arose in your mind for further reflection and consideration? What else did you notice during your reading?

Continuing. Continue your change journaling practice as you read further in the field manual and as you apply some of the practices suggested.

Dip or Dive

1. *A Whole New Engineer. Dip.* Peruse the website *www.wholenewengineer.org*. Watch a video, read the *Big Beacon Manifesto*, or otherwise skim or scan other resources available there. In what ways do the resources you've watched, skimmed, or scanned resonate with the material of this chapter? In what ways do they resonate with or contradict your experience? What else do you notice? *Dive.* Get the book, and skim, scan, or read it and journal about salient ideas that you find interesting, surprising, confirming, or infuriating.

2. **Non-traditional educational program.** *Dip.* Consider a non-traditional higher educational program that interests you (*www.minerva.edu, www.outlier.org, www.coursera.org, www.udacity.com, www.edx.org*, or some other non-traditional program you are curious about). Go to the website and peruse the architecture, range,

manner of offerings, assessment, and experience of these programs. In what ways do these programs pose a threat or not to traditional educational offerings? *Dive.* Sign up for a course from one of these offerings and compare and contrast the experience to offerings at your institution.

3. **Effectuation or little bets.** *Dip.* The idea of being effectual (Sarasvathy, 2008) or making little bets (Sims, 2011) is explored in two *Big Beacon Radio* podcast episodes (Goldberg, 2016, May 9; 2016, November 6), *https://www.voiceamerica.com/episode/92149/ effectual-entrepreneurship-an-interview-with-saras-d-sarasvathy* and *https://www.voiceamerica.com/episode/95621/little-bets-and-breakthrough-ideas-an-interview-with-peter-sims*. Listen to one of them and take notes on key ideas and insights. What key ideas do you take away from your listening? *Dive.* Get one of the books and skim, scan, or read it. Journal as you do this and contrast these ideas with the normal notion of planning.

Reflectionaire

1. **Empowering educational experience.** Consider an educational experience that was empowering or in some way enriched your sense of competence and self-confidence. What elements of the experience contributed to these feelings? To what extent did a particular teacher, learning materials, performance, outcome, subject matter, or other factors come into play? To what extent was the experience relaxing or stressful? To what extent were there obstacles involved? What was the trajectory of your mood during the experience? What else do you notice now about the experience?

2. **Disempowering educational experience.** Consider an educational experience that was either disempowering or in some way diminished your sense of competence and self-confidence. What elements of the experience contributed to these feelings? To what extent did a particular teacher, learning materials, performance, outcome, subject matter, or other factors come into play? To what

extent was the experience relaxing or stressful? To what extent were there obstacles involved? What was the trajectory of your mood during the experience? What else do you notice about the experience?

3. **Student responsiveness to a teacher [for teachers].** As a teacher, which of your classes do your students respond to most positively? Which of your classes do students respond to least positively? What feedback have you received about these courses that helps explain the difference? What theories do you have to explain the difference? What is it about the material or courses themselves that helps explain the difference? What is your attitude to the different courses? To what extent are the courses structured and to what extent do the students get to exercise choice? What other differences do you notice between the courses that helps to explain the difference?

4. **Student responsiveness to a teacher [for students].** As a student, which of your classes do you respond to most positively? Which of your classes do you respond to least positively? What feedback would you give your teachers that would help them understand student engagement? What is it about the material or courses themselves that helps explain the difference? What is your attitude to the different courses? To what extent are the courses structured and to what extent do the students get to exercise choice? What other differences do you notice between the courses that helps to explain the difference?

5. **Students today are not . . . [for faculty].** Sometimes faculty members assert that students today are different from students of yesteryear. With or without a colleague or group of colleagues, quickly make 5-15 responses to the following prompt: "Students today are not . . ." [fill in the blanks]. Following this exercise, consider asking some students to do exercise 6 as a companion.

6. **Teachers today are not . . . [for students].** Sometimes students complain about their teachers. With or without a colleague or group of colleagues, quickly make 5-15 responses to the following prompt: "Teachers (instructors) today are not . . ." [fill in the blanks].

Following this exercise, consider asking some teachers to do exercise 5 as a companion.

7. **Students and their passionate pursuits [for teachers or students].** Think of times when you have seen students spending large amounts of time passionately pursuing some interest. Sometimes this may be in class, but often it is some extracurricular activity. Make a list of those passionate pursuits that you are aware of. What is the difference between students who pursue their interests passionately or not?

Little Bets

1. **Reading group for the *Field Manual*.** Form a reading group comprised of faculty, staff, and students interested in more engaging and effective education. Read *A Field Manual for a Whole New Education* chapter by chapter, discussing each chapter and ways in which the material in the chapter is or is not relevant to the change circumstances of your school. Look for actionable little bets whereby small experiments can be undertaken to both gather data and explore change paths.

2. **Reading group for other resources.** Form a reading group comprised of faculty, staff, and students interested in more engaging and effective education and choose one or more of the key resources cited in *A Field Manual for a Whole New Education*, discussing each resource and ways in which the material is or is not relevant to the change circumstances of your school. Look for actionable little bets whereby small experiments can be undertaken to both gather data and explore change paths.

3. **I like, I wish exercise.** Gather a small group of stakeholders (students, faculty, staff, employers, alumni, etc.) affiliated with your education program virtually or in person. It is best to do different classes of stakeholder separately (students with students, alums with alums, etc.). Tell them that you are trying to make educational improvements and ask them to take sticky notes or small pieces of paper or

cards and have them write out "I like . . ." where they fill in the blank regarding things that they like about your program. Have them fill out between 2-5 sticky notes per person. Gather those up after they are done and post them to a central board. Thereafter have them fill out "I wish . . ." where they fill in the blank regarding things that they wished were different about the program. Again, have them fill out between 2-5 sticky notes per person. Gather those up after they are done and post them to a central board. Save and organize the responses, as this can be a very helpful source of data for change making later.

2

A Whole New Engineer in 4.8 Minutes

At its core, *A Whole New Engineer* (Goldberg & Somerville, 2014; hereafter WNE) is a lessons-learned memoir of contrasting change at two very different institutions, Franklin W. Olin College of Engineering in Needham, MA, and the University of Illinois at Urbana-Champaign. In many ways, the schools couldn't have been more different; in WNE we called them the (collaborative) David and Goliath of engineering education. Olin was at the time of the telling, a start-up experiment in bringing about a more engaging, holistic engineering education in liberal arts college format. Illinois, on the other hand, was at the time an old, large, highly rated public land-grant institution with faculty hired for their research acumen and a large undergraduate engineering program that had little appetite for extensive curriculum or culture reform. The two schools were brought together by an improbable series of events initiated in 2006 that resulted in a formal partnership in 2008, the Olin-Illinois Partnership. Although many at Olin thought they had already found the "big it" of engineering education reform on their own, and many at Illinois couldn't see any benefit in partnering with a school without a significant research footprint, neither was prepared for the deep insight and learning that would be revealed by the active reflection on the contrasts between two such different schools and cultures.

To understand the power of the stark contrasting case studies, we recommend going back and reading the first two chapters of *A Whole New Engineer* (WNE, pp. 1-68) in which the change stories of Olin and Illinois are told side by side. Many people react to the Olin story by flippantly suggesting, that they, too, could create something as cool and innovative as Olin if only they had a half-billion-dollar philanthropic gift to build new buildings, hire new faculty, and create a new curriculum. Of course, these remarks underestimate the complexity and quality of the accomplishment, but side-by-side with the Illinois story, the flippancy also highlights a key misunderstanding: the idea that the money, the new buildings, and a new faculty were the core constituents of the kind of change that occurred.

The Illinois story starts with college leadership and a faculty reluctant to make any significant change in undergraduate education at all, with no new buildings for educational infrastructure, and almost no new money for curricular innovation. But the surprising thing then and now, was that a small pilot project—so small we called it the *peashooter*—in the iFoundry (Illinois Foundry for Innovation for Engineering Education) incubator was able to bring about significant emotional and cultural change that was surprisingly strong and qualitatively similar to the changes that had occurred at Olin.

For our purposes here, the key findings of *A Whole New Engineer* are as follows:

- **Student unleashing** is central to educational transformation in our creative era.
- Almost all key change variables in student unleashing are **emotional** and **cultural.**
- PhD-trained faculty trained need **coaching skills** to facilitate student unleashing.
- **Culture** plays a key role in both resisting proposed changes and sustaining effective changes.

- Modern theories and practices of **motivation, mindset,** and **multiple intelligences** are helpful in nudging traditional educational cultures in useful ways.
- Effective **organizational change theories & processes** can be applied to accelerate a whole new education.

In this chapter, each of these is briefly discussed in turn.

Reflexercise 2.1: Read Chapters 1 and 2 of *A Whole New Engineer*. As you read the chapters make notes of key terms, salient experiences, interesting outcomes, and other points that come to mind.

Reflection. Compare and contrast the Olin experience to the Illinois experience. Compare and contrast the Olin and Illinois experiences to your own experiences with educational change efforts. In what ways are you inspired or discouraged from the change-making stories in these chapters? What were the key focal points of change in these stories? What else do you notice in reflecting on this reading?

Unleashing as Key to Education Transformation

In WNE, student unleashing is a bilateral process by which faculty let go of strict authority over the learning environment and students feel free to and do embrace and create learning experiences with strong passionate engagement. In creating a new school, Olin had built student unleashing into their buildings, faculty hiring, and classroom experiences (WNE, pp. 1-31), but the importance and centrality of unleashing came into sharper focus through two experiences (WNE, pp. 33-68).

On a trip to Olin in 2008, a team from the University of Illinois visited Olin's campus and observed that even as early as freshman year, students at Olin shifted in identity from being mere high school students to embracing the emotional-cultural core of passionate makers and engineers. In the

17

book, we called this the *Olin effect* and labeled this shift as a key goal of educational change. This might not seem like a big deal, but the 2008 observation of the Olin effect put the Illinois team on the lookout for the right stuff, that is, emotional-cultural, not just intellectual, change. Later, in the 2009 freshman pilot of iFoundry, we observed a qualitative shift toward student autonomy when halfway through the semester a student commented during a *kaizen* or improvement session (WNE, p. 63)

> "We weren't sure you were serious about us doing what we wanted
> to do, and then we realized you were, and it was really cool."

A burst in student energy and engagement following that point in the semester helped us understand the power and low cost of student unleashing. New buildings, new faculty, and lots of money are less important than a laser focus on making sure there are unleashing experiences throughout an education.

The Five Pillars of Unleashing are Emotional & Cultural

In the sensemaking journey of Olin and Illinois described in WNE, we listed and discussed the five pillars of unleashing (WNE, pp. 125-148) as follows:

1. joy
2. trust
3. courage
4. openness
5. connectedness, collaboration, and community

It is useful to think of these words as forming an interlinked system for unleashing. Items 1, 4 & 5—joy, openness, connectedness, collaboration, and community—operate at the level of the group to create a backdrop for unleashing. A joyful mood creates an atmosphere of positive emotion and energy from which other good things can flow. An openness to new ideas, people, and action creates an encouraging space for failure, success, growth, and learning. An emphasis on relationships and connection to others is a social backdrop to students and faculty working together productively, which helps to create a sustained community with an identifiable culture.

Item 2, trust, packs an important wallop in the system. By one account (Brothers, 2005), a person is considered trustworthy if they are sincere, reliable, and competent, but trusting a person goes beyond a historical assessment of a person's performance. Saying, "I trust you," and backing up those words with appropriate behavior, can change the person trusted in ways that can elevate their performance in surprising ways.

Courage is the topper of this interlinked unleashing system. A person trusted within a joyful, open environment, of helpful collaboration and community, finds it easier to have the courage to be creative and take action to try new things, with less direction or fear of failure. Having students experience this kind of courage in the classroom can be an important forerunner to them having the courage to take important, creative, or difficult action later in life.

Together these five items form a powerful system for unleashing student potential to the possibilities in their lives. Not all of the five items are necessary in an educational system for unleashing, and the most capable students figure out how to unleash themselves against a host of obstacles. Nonetheless, if education itself is to cultivate insightful graduates expeditiously and well, it must increasingly support their systematic unleashing and development in ways like those described in WNE

> **Reflexercise 2.2:** Consider a time in your life when you were *unleashed* (in the sense of this chapter) to do some difficult job or task with little or minimal guidance.
>
> **Reflection:** What was the trajectory of your thoughts and feelings as you started and completed that job or task? Which of the five pillars of unleashing were relevant or not to your experience? What kind of role did your boss or supervisor play in your unleashing? What else do you notice about this personal unleashing experience?

PhD-Degreed Faculty Need New Skills to Unleash Students

Together these five items—joy, trust, courage, openness, and connectedness, collaboration & community—form a powerful system for unleashing students to the possibilities in their lives, but there's a hidden problem. Although the PhD can be a powerful unleashing experience for those who obtain one, it isn't a systematic teacher of how to unleash others; most faculty members in higher education are trained in narrow research specialties at the PhD level, but mastery of a narrow field doesn't necessarily teach the skills necessary to help others unleash. The need to learn lots of the well vetted, largely theoretical knowledge in narrow fields serves the goal of perpetuating those narrow research disciplines at the same time it implicitly creates a culture wherein the job of teaching is merely to pass on what is known to the relatively empty vessels called students. Of course, this approach is not without value and goes back to the very inception of the university in 1088 as an assembly of experts (WNE, pp. 75-76), but as mere theoretical knowledge and lectures on that knowledge are widely accessible and fairly freely available on the web, returns to expertise have diminished (WNE, pp. 76-91) especially in teaching, and it becomes increasingly important to attend to students and their ability to explore, create, collaborate, and lead.

The dichotomy between the expert knowing what another should know (the expert teacher) and the guidance skills to unleash others (the unleashing coach) are somewhat in conflict as a mere matter of logic. If someone is to be unleashed, what should they do, what should they learn, what should they be required to know? Disciplinary requirements in part give some of the answer, but there must be some *domain of student autonomy* that is respected by all concerned, and then if the autonomous human-in-training is to be supported in some way, how can that be done in a manner that not only respects autonomy, but encourages its blossoming and development?

Fortunately, the private sector has helped solve this problem with the widespread adoption of executive coaching. Increasingly C-suite members (CEO, CTO, CFO, CMO, etc.) have personal executive coaches to help them learn to communicate, grow, and develop as leaders. The idea put forth in WNE (pp. 175-189) is that the professor of our times must

(1) be knowledgeable in his or her discipline as before and (2) have a working knowledge of the rudimentary skills of the executive coach. In WNE, we called out five skills in particular: (1) noticing, (2) listening, (3) questioning, (4) speech acts, and (5) understanding and reframing stories. This is a good starter set, and later we will both double down on these skills and elaborate on them in important ways.

Culture as a Problem, a Solution & a Reluctant "Student" to be Educated

While much education change effort is directed at content, curriculum, and pedagogy—and indeed these are important artifacts of culture—WNE (pp. 191-207) calls out culture itself as a primary locus of change. When new initiatives are resisted, it is culture that cries out, "that's not how we do it here," (Kotter & Rathgeber, 2016) through the individual voices of myriad students, faculty, alumni, and other stakeholders enculturated by the current system, voices who are comforted by the buildings occupied, the stories told, the roles filled, and the rituals and processes practiced. Interestingly, it is also culture that preserves and sustains the changes made during effective change initiatives, long past the active participation of those who instigated change in the first place, and long past the time when the original reasons for change have been forgotten.

As a result of the awareness of the crucial role of culture in both resisting and sustaining change, WNE emphasized understanding culture and using that understanding to shift key cultural features (WNE, pp. 191-207). There are different theories of culture, but one that has been particularly helpful in WNE is Schein's theory of organizational culture in which he defined culture as follows (Schein, 2009, Location, 557-559):

> Culture is a pattern of shared tacit assumptions that was learned by a group as it solved its problems of external adaptation and internal integration, that has worked well enough to be considered valid and, therefore, to be taught to new members as the correct way to perceive and feel in relation to those problems.

He also defined three levels of culture as follows:

1. Artifacts
2. Espoused values
3. Underlying assumptions

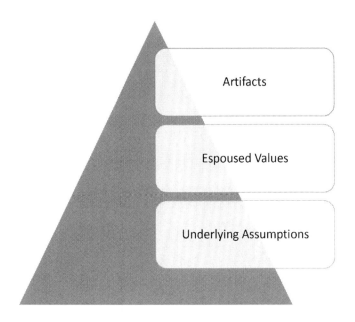

Figure 2.1: Schein's model of culture (2009) has three levels:
Artifacts, espoused values, and underlying assumptions.

This decomposition can be particularly helpful to having students and faculty examine the current culture of a department through the field glasses of an anthropologist. Starting with artifacts, the hunt can begin for physical and institutional artifacts such as courses, classrooms, other physical spaces, textbooks and other learning materials, roles and hierarchies, regular events, rituals, and other structures that are the backdrop of departmental activity.

Espoused values may then be uncovered explicitly or implicitly in the language, documents, web pages, advertising copy, and common recruiting and onboarding stories that are told about what is important to the culture (and what is off limits or tabu).

Underlying assumptions are often difficult to identify explicitly because they are so deeply believed and so widely accepted, they are barely noticed, but it is this level that often causes the most resistance to what otherwise would seem to be an innocuous proposed change.

An existing culture is difficult to change directly, but it can be nudged by (1) studying it like a field anthropologist (2) identifying key leverage points for change, (3) being curious about the historical reasons why the culture evolved the way it did, (4) tapping into the underlying assumptions of the culture by recognizing the legitimacy of that evolution, and (5) telling different stories about the historical forces that are acting today and why change is necessary, while (6) preserving the best of the old culture that is still working well. Additionally, a number of social science theories and practices are particularly helpful in bringing about the student unleashing and the culture in which it lives.

Interestingly, new cultures can be cultivated quite easily and directly within small groups of voluntary participants. At Olin, in iFoundry, in new change initiatives, and in workshops and conferences devoted to educational change, Mark and Dave have both noticed the ease and speed with which a new set of artifacts, espoused values, and underlying assumptions can emerge. Moreover, this nascent culture formation is often accompanied by a fundamental shift toward a good mood. Later in the field manual, we will exploit this difference between the resistance of a large extant culture and the malleability of a small emergent culture to help accelerate change through what we call affective implementation.

Reflexercise 2.3: Consider two organizations you have been a part of. Use Schein's model—artifacts, espoused values, and underlying assumptions—and make a list of these elements for each of the two organizations.

Reflection: Compare and contrast key similarities and differences in the cultural elements of the 2 organizations. In what ways do these similarities and differences in culture reflect similarities and differences in your work experiences in these two organizations?

3 Elements for Nudging Educational Culture

For educational culture's specifically, WNE advocated three theories of modern social science as particularly helpful in nudging educational culture:

1. Multiple intelligences (WNE, pp. 93-95)
2. Mindset (WNE, pp. 96-97)
3. Motivation (WNE, pp. 149-174)

Howard Gardner's work (Gardner, 1993) redirected educational emphasis on cognitive IQ to the existence and human value of multiple intelligences: musical, linguistic, spatial, bodily-kinesthetic, logical-mathematical, interpersonal, and intrapersonal. Taking a cue from this work, WNE (pp. 100-116) lists the *six minds of the whole new engineer* as follows:

- Analytical mind
- Design mind
- Linguistic mind
- People mind
- Body mind
- Mindful mind

More information is available on each of these in the book, but the key point for this field manual is that different quarters of higher and professional education might choose different naming conventions and slices through Gardner's multiple intelligences than did WNE, but the key idea is to recognize a breadth of qualitatively different aptitudes, skills, and practices in thinking about the core subject matter emphasized in any educational reform effort.

Recognizing multiple intelligences as a more holistic approach to educational content is an important step, but in the Chapter 4 of WNE, we also called out the importance of one's *belief about intelligence* to a student's education, growth, and development as important. Carol Dweck (2006) made and researched an important distinction between what she termed *fixed* and *growth* mindset. In Dweck's terms, if one has a fixed mindset, one believes that character, intelligence, creativity, and other capabilities are static and given; essentially, they are predetermined and can't be changed in any significant way by one's behavior. In this view, education is mainly about revealing who's got the right stuff, and schools sort and slot people into different roles in the world of work according to their revealed capacities. In the fixed mindset story, one should stick to one's knitting, and avoid the risks of revealing what one isn't very good at.

On the other hand, if one has a growth mindset, one believes that one's own effort is a major determinate in one's character, intelligence, creativity, and other capabilities. In this view, education encourages everyone, regardless of current ability, to pursue personal development, growth, and learning. In the growth mindset story, one works by developing strengths and weaknesses sufficiently, recognizing failures and setbacks as part of the process.

Earlier we spoke of focusing on the individual faculty member and encouraging the shift from purveyor of expert knowledge—the sage on the stage—to the coach who can support student unleashing. Just now we spoke of two shifts--the shift from IQ to multiple intelligences and the shift from fixed to growth mindset—that help us reflect on deep cultural assumptions within educational institutions that can support appropriate educational change. Here we direct our attention to students and their *motivation* and *engagement* by examining some of the key tenets of modern motivational theory. Chapter 5 (WNE, pp. 117-174) started by recounting Edward Deci's groundbreaking experiments that showed how rewards can actually demotivate individuals to perform tasks that otherwise might be fun or intrinsically rewarding.

This led to a discussion of Deci & Ryan's (2018) *self-determination theory* (SDT), and Dan Pink's popularization of these and other ideas in his influential text *Drive* (Pink, 2009). We highlighted these with the triple: (1) autonomy, (2) purpose/connectedness, and (3) mastery. More can be found in WNE, but here we emphasize the connection between these things and the five pillars of unleashing discussed earlier. While traditional education has focused on obedience to authority, allegiance to disciplinary expertise, individual performance, and disciplinary competence as something that requires external assessment, a whole new education balances these traditional values with a greater respect for individual autonomy, a broader range of purposive or meaning-making activities, the importance of community and connection, and mastery and development as something that is often its own reward.

Beyond Committees: Toward More Effective Change Practices

The change agents at Olin and Illinois iFoundry recognized that if substantive change was to take place at either institution, there would have to be greater attention to the *processes of change* adopted. In the case of Olin, there was nothing at the beginning of the school, so new faculty and staff were hired, and they started from a blank piece of paper as described in Chapter 1 of WNE. A greenfield effort is expected to do something different, but the tug of traditional educational culture can still constrain even brand-new efforts to be less innovative than they had hoped. A key and in-part accidental way in which Olin disrupted the siren song of the status quo was partnering with students to pilot and design the school's curriculum in the *partners year*. (WNE, pp. 17-18).

At Illinois, the change effort described started off in pairwise conversation between two faculty members, and an early offer from the Dean of Engineering to form a traditional committee to study change was rejected (WNE, pp. 37-38), because, generally speaking, committees are designed to support the status quo and sort through the manifold reasons why something new can't be done and why this or that can't be changed. The conversation grew to become a larger team of engaged faculty and students, and that informal team branded itself as the iFoundry incubator, which then became a college-sponsored effort of the same name. The rejection of the traditional bureaucratic path to making change through a committee was essential to achieving something special.

To understand different paths to substantive change, chapter 9 of WNE (pp. 209-230) told several change stories beyond those of Olin and iFoundry at Illinois. It then compared and contrasted two approaches in the change literature. The first was the process-centered approach to change of Harvard's John Kotter (1996). In Kotter's telling, change is a step-by-step process initiated by organization leadership to engage organization membership in a systematic way. The second was the human-centered approach to change of the Heath brothers (Heath & Heath, 2010). In their telling, change can be driven from bottom, top, or middle, but they argued that change must appeal to both the heart (the "elephant") and the head (the "rider") at the same time it clears obstacles out of the way of those asked to change (the "path").

Later in this field manual, we will combine process- and human-centered approaches into a single step-by-step approach that brings the systematic sequencing of a step-by-step approach with broad-spectrum support of head, heart, and obstacle clearing. Doing so will permit substantive changes such as major curriculum revision in shorter time periods than was previously believed possible.

More can be said about the lessons of WNE, and those interested in learning more should consult the book directly. The next chapter turns to summarizing some of the practical lessons we've learned since the book's writing in 2014.

Reflexercise 2.4: Journal or make some notes about your takeaways and action-aways from this chapter.

Some questions. Skim back over this chapter. What parts resonated with you? What parts did you disagree or have some concern with? What terms were used in ways that seemed familiar or unfamiliar? What experiences in your life connected to what you read? What questions arose in your mind for further reflection and consideration? What inspired you to action? What else did you notice?

Dip or Dive

1. **Motivation.** *Dip.* Watch the short video (RSA Animate, 2010, April 1) *https://www.youtube.com/watch?v=u6XAPnuFjJc* adapted from a talk by Dan Pink summarizing the findings of his book *Drive* (Pink, 2009). In what ways are the ideas of motivation discussed in this short video relevant (or not) to the reform of education? What points are most interesting to you? What else do you notice as you watch this video? *Dive.* Read the book and journal about key points that you find interesting, distressing, or motivating.

2. **Mindset, fixed & growth.** *Dip.* Watch this short video (*https://www.youtube.com/watch?v=hiiEeMN7vbQ*) by Carol Dweck (2014), author of *Mindset* (2006) in which she distinguishes between a growth mindset and a fixed mindset. In what ways are the ideas of mindset discussed in this short video relevant (or not) to the reform of education? What points are most interesting to you? What else do you notice as you watch this video? *Dive.* Read the book and journal about points that seem relevant to educational reform.

3. **Multiple intelligences.** *Dip.* Watch Howard Gardner discuss the ideas from Multiple Intelligences (Gardner, 1993) on YouTube (Gardner, 2009) *https://www.youtube.com/watch?v=l2QtSbP4FRg*. What points are most interesting to you? What else do you notice as you watch this video? *Dive.* Skim, scan, or read the book and journal about salient items as you go through it.

4. **Power of vulnerability.** Watch Brené Brown's (2010) TED talk (*https://www.ted.com/talks/brene_brown_the_power_of_vulnerability?language=en*). In what ways are these ideas connected to the five pillars of unleashing? In what ways do Dr. Brown's insecurities connect to your own? What else do you notice about this video? *Dive.* Read the book *The Gifts of Imperfection* (Brown, 2010) and journal about points that connect to the reform of education. You may substitute any of Brené Brown's other books if you prefer.

5. **Change leadership.** *Dip.* Use the internet to find summaries of *Switch* (Heath & Heath, 2010) and *Leading Change* (Kotter, 1996).

Compare and contrast the approaches to change advocated in the two books. What are the advantages and disadvantages of each for making change in an educational environment? What thoughts do you have about how you would combine the best of both approaches to make rapidly embraced, effective change in your school? What else do you notice? *Dive.* Repeat the exercise using the raw books as your source material.

Reflectionaire

1. **New school or program.** If you are considering starting a new school or a new educational program within an existing school, read or reread Chapter 1, "Engineering Happiness: The Olin Experience" (WNE, pp. 1-31). As you read the chapter, make note of some of the elements of the founding of Olin that seem useful to creating an engaging educational experience. What elements can be used directly in your program? On elements that cannot be used directly, what modifications or other framings can be made to render them useful to your program? What else do you notice in reading the chapter?

2. **Legacy program.** Making change in an existing or legacy program tends to be more difficult that starting a new program due to cultural resistance surrounding the existing program. If you are considering changing a legacy or existing program, read or reread Chapter 2, "The Incubator: Helping a Big Old Dog Learn New Tricks" (WNE, pp, 33-68). As you read the chapter, make note of some of the elements of the founding of the iFoundry incubator at Illinois that seem useful to transforming an existing educational experience. What elements can be used directly in your program? On elements that cannot be used directly, what modifications or other framings can be made to render them useful to your program? What else do you notice in reading the chapter?

3. **Student unleashing experience.** Consider a type of experience at your school in which students become unleashed in the sense of this chapter. What is it about this experience that leads to unleashing? In what ways do the five pillars play a role in student unleashing? What else do you notice about the experience?

4. **Committee experiences.** Reflect upon the different kinds of committees you've been on. Compare and contrast the positive versus negative committee experiences you've had. What factors are associated with positive experiences? What factors are associated with negative experiences? What kinds of things do committees appear to do well? What kinds of things do committees do poorly? What else do you notice about committee work?

5. **DIY (do it yourself).** Upon reading this chapter, what concept, experience, innovation, or other object caught your attention that deserves further reflection? What is it about this object of your attention that interests you? What have you experienced in connection with this object? What other questions need to be asked and reflected upon in connection with your DIY reflectionaire? Reflect on those questions. What else do you notice as you reflect upon this object?

Little Bets

1. **Artifact hunt.** In a small group go on an "artifact hunt," which is similar to a scavenger hunt, but the object is to make a list of artifacts (spaces, signage, literature, symbols, hierarchy, rituals, etc.) in your organization's educational culture. Divide the small group into pairs or triples armed with clipboards and smart phones and have them make lists, record images, and records sounds of artifacts in the educational culture. Consider the function of these artifacts and some of the larger messages they send about educational culture you are observing.

2. **Espoused value hunt.** Reflect on the culture and mood of an existing educational culture with which you are familiar. What are some of the key artifacts of that culture (spaces, signage, literature, symbols,

hierarchy, rituals, etc.). When people speak of the institution, what do they say about it? What words are used in expressing its values? What are some of the deepest assumptions of the organization that are so ingrained that they don't need to be said? What is the general emotional mood of the program? What else do you notice?

3. **Underlying assumption hunt.** Consider the results from items 1 and 2 above and make a list of conjectures about deep unspoken, underlying assumptions from which they may flow. Next, read Carl Rogers's remarkable (and originally rejected) paper *Current assumptions in graduate education: A passionate statement* (Rogers, 1969, pp. 168-187) or simply go through a list of the assumptions here: *http://www.learning-knowledge.com/teaching.html*. In your view, which of Rogers's assumptions apply to your educational institution? What else do you notice as you read these assumptions?

4. **Visit an innovative education program.** Visit Olin College or an educational institution, in-person or virtually, that you consider to be an educational innovator relevant to your discipline. What do you notice programmatically? What do you notice culturally and emotionally? What stories, practices, and artifacts are worthy of possible imitation or modification at your institution? What stories, practices, and artifacts are problematic at your school? What else do you notice?

5. **DIY little bet.** What little bet needs to be made as a result of your thinking about the material in the book so far? What is easiest to try or do with existing means? What is the smallest possible experiment that will get you some interesting results? What do you notice from doing this experiment? What else needs to be done?

3

New Lessons from the Front Lines of Change

What Have We Learned Since 2014?

Since the publication of WNE, Mark Somerville has continued as a Professor at Olin College and has played a leadership role as Dean of Faculty and Provost; he has also consulted with change initiatives around the world. Dave Goldberg has continued as President of ThreeJoy Associates (a for-profit consulting, coaching, speaking, training, and facilitation firm) and Big Beacon (a non-profit organization for the transformation of engineering education), and continues to give talks, facilitate workshops, coach change leaders, and help bring change to higher education around the globe. In these roles, Mark and Dave continue to learn how to think about and practice change in higher education. The following is a short list of some of the key things we've learned since the publication of WNE:

- **Grokking the digital gaggle.** In WNE, we broadly spoke of the need for change in education due to the three missed revolutions: the entrepreneurial evolution, the quality revolution, and the information technology revolution. In this book, we focus on the IT revolution and the rise of the digital gaggle—the increasingly powerful combo of digital technology, AI, machine learning, and robotics. At the same time the methods of WNE have helped others unleash students and make meaningful educational change, they also appear to give us a fresh, practical approach toward gaggle-proofing higher education and preparing students for lifelong human insight.

- **Considering new business models.** When we wrote WNE, the financial pressures on higher education were certainly present, but they have become more pronounced in recent years. Integrating the lessons of the book with a questioning of current transactional *learn-then-earn* model can lead, we think, to new educational models that improve learning quality and create societal benefits in ways previously not considered possible.

- **Adopting rigor in a core basis set of "shift skills."** In the original text, we emphasized what others like to call soft skills, but increasingly we believe that there is a rigorous basis set of what we call shift skills or insight shifts that should be part and parcel of every student's (and every higher educator's) tool kit.

- **Thinking with co-contraries.** We've increasingly emphasized co-contraries (sometimes called *polarities*) or opposites that need each other (Elbow, 1987; Johnson, 2014, 2020) as a useful framework for capturing and designing around the complexity of educational change.

- **Challenging the privilege of theory over practice.** For some time in a university education, theoretical understanding has been privileged over practical ways of knowing. Increasingly, we've come to understand such *theory privilege* as an obstacle to doing the right things in engineering education and higher education more generally.

- **Accomplishing change in months, not years.** In WNE, we thought that curriculum and institutional change in higher education was necessarily a multi-year undertaking, but we now increasingly believe that proper process and facilitation together with a focus on key cultural and emotional variables can lead to major shifts in curriculum and culture in months not years.

- **Affective implementation carries over the mood of change and the plan.** A key lesson of WNE was the importance of culture and emotion to effective change, and one of the things we've learned since 2014 is to carry over the culture and emotion of effective change processes as well as the change plan the process creates.

Each is discussed in the remainder of the chapter in turn.

Grokking[1] the Digital Gaggle

In WNE, we coined the phrase *the missed revolutions* and enumerated three of them: (1) the entrepreneurial revolution, (2) the quality revolution, and (3) the information technology (IT) revolution, and we used the term "missed" in the sense that the methods of the revolutions are practiced by industry and have been largely missed by universities. Yes, various university departments teach entrepreneurship, quality methods, and IT, but universities as institutions haven't typically integrated the lessons of the missed revolutions into their core business.

One of the things we've reflected on since the publication of WNE is where the IT revolution is heading and how it interacts with the lessons of that book and this manual. Of course, COVID-19 gave universities and colleges around the world a wake-up call in 2020 and 2021 on using digital technology in the delivery of education remotely, but here our concern goes deeper than that necessity becoming the mother of that limited innovation. At the beginning of the manual, we introduced the term "digital gaggle" as a stand-in for the aggregate of digital technology, artificial intelligence, machine learning, and robotics. Given that we are both engineers, some might expect us to sing the praises and sow the seeds of educational technology (EdTech) or extoll the not insubstantial virtues of the digital gaggle, yet here we head in the opposite direction. Our job is to understand

---

1. The term "grok" appears first in Robert Heinlein's (1961) sci-fi novel *Stranger in a Strange Land* and is a fictional Martian term. The term now appears in dictionaries and its meaning is debated by critics. For us, in the context of the manual, to grok is to understand deeply in head, heart, and body.

where the digital gaggle is heading, not to amplify it; the gaggle will grow without our cheerleading. Instead, we wish to stand back from the gaggle's effects, understand the twists and the turns of its development, so that we can better understand what kinds of skills and mindsets are particularly human in a world where the gaggle takes on more and more routine and high-end knowledge work. We will interrelate our processes and the gaggle throughout the manual, but here we start with a short examination of the long march of the digital gaggle.

A Short, Incomplete Survey of the Long March of the Gaggle

Our historical survey of the gaggle will be as unsatisfying as some of the genealogies in the bible, where so and so begat so and so who begat so on and so forth, leaving us at some current circumstance that gets a somewhat fuller narrative. Nonetheless, one place to start our story is in the late 19th and early 20th centuries with widespread electrification. Electrification begat electric lighting which begat electric appliances, telephones, radios, phonographs, and then televisions. Relays begat vacuum tubes begat transistors, which in a variety of stages begat digital computers. Hard-wired programs (literally with patches and plugs) led to instructions in memory, then assembler and high-level languages entered by punch card, tape, and terminal.

Early cryptographic, business, scientific, and engineering computing led to dreams of artificial intelligence, though artificial checkers and planning programs (Feigenbaum & Feldman, 1963) and natural means (Holland, 1975) such as neural networks and evolutionary and genetic computations. Magnetic core memory and magnetic tapes begat spinning disks and later solid-state memories and storage devices. Time-shared computers were enabled by powerful operating systems driven by increasingly sophisticated command languages. Wired hardware begat printed circuit boards which begat integrated circuits which led to the pivotal microprocessor. Microprocessors put not-so-powerful, then incredibly powerful, calculators at human fingertips and then in short order, personal computers, and server farms.

Increasingly personal and other computers were connected to one another by modems and then directly using standardized network protocols and interfaces. At about the same time cellular phones arose, moving from analog to digital quickly, then quickly outfitted with cameras and mp3 players then texting capability and other functions not typically thought of as being phone-like. Widespread networks led to email, bulletin boards, Gopher, and to internet browsers, and these things started to appear on personal computers and phones both.

The browser begat e-commerce and widespread copying and sharing of text, audio, and video, whether it was someone's intellectual property or not. Disks and servers got together with sophisticated operating systems in what we now recognize as cloud computing. Phones and computers converged into the smartphone, and this led to smartphone apps, which increasingly paved the way for the efficient transmission of all the foregoing stuff, together with an increasing power of distributed networks of people and devices in ridesharing, home sharing, X-sharing, and gigging apps in ways that continue to ripple through the global economy.

These networks also wiggled their way into banking, markets, finances, and ultimately into currencies independent of any central banker. Along the way manufacturing of things themselves became more and more the realm of the gaggle with computer-aided design and manufacturing software driving increasingly sophisticated machine tools and robots. Robots themselves were unleashed inside our homes, offices, and warehouses, and more and more physical labor is being carried out by elements of the gaggle. Earlier dreams of artificial intelligence returned with a vengeance with graphics processing units (GPUs) powering advanced machine learning, with human performance increasingly rivaled by statistical learning methods, deep learning neural networks and genetic algorithms and programming. As we write these words, these increasingly smart technologies are being integrated into apps, smartphones, and personal and business computers in ways that increase their power, reach, and effect.

A Change in Interconnectedness, Closeness, Interruption, and Augmentation

We recount the long march of the digital gaggle first, because some younger readers may assume that the gaggle has always been with us. For those of us who have lived through the emergence of many of its elements, it is a story of astounding inventiveness and innovation. And to be sure, we are both men of the gaggle, Mark as a semiconductor engineer and scientist, Dave as an AI researcher (Goldberg, 1989, 2002). We respect the ways in which the pieces of the gaggle puzzle have been assembled step by painstaking step, and we admire the boldness and daring of gaggle researchers, engineers, and businesspeople alike. In this way, the gaggle arose from humble beginnings to become increasingly functional and powerful forces in our lives.

As we reflect on this scene, we also notice some qualitative changes in the way the gaggle affects us as human beings. In the story of the gaggle itself, we note increasing *interconnection* over time between the tech components of the gaggle and the humans with which the gaggle interacts. At first, human interaction first was limited to technical specialists, but as time went on, it became easier or ordinary humans to be in the loop, both physically through terminals and touch screens, and in terms of ease of interaction through software that could be directed through mouse clicks and touch.

Likewise, there has been increasing *closeness* of the gaggle with humans. At first, gaggle components were locked away in rooms with restricted access, but increasingly components have moved from far away rooms to phones, watches, automobiles, refrigerators, and other devices, one or more of which is almost always within our reach.

As the gaggle has become more interconnected and closer to the humans with which it interacts, it has found ways to *interrupt* us more and more. In the old days, a computer would sit in its far away air-conditioned and locked room, and the only way it might interact with a distant human would be through the postal service with a utility bill. Now, with devices everywhere, the gaggle stands at the ready to interrupt us, distract us, and in many ways keep us from one another. Given that much of what animates the gaggle is the making of a profit, the more we are interrupted, the more money there is to be made.

Finally, the rise of the gaggle has led to gaggle members augmenting human work, both quantitatively and qualitatively to the point where gaggle members augment and sometimes displace human workers. Run-of-the-mill automation of repetitive manual tasks is an old story, but the replacement of high-end knowledge workers by AI and ML is relatively new, and it will continue unabated as artificial intelligence and machine learning are increasingly included in all the gizmos and facets of our lives.

Reflexercise 3.1: What is your favorite app, program, website, device, or gizmo in the digital gaggle? What is your most annoying app, program, website, device, or gizmo in the digital gaggle?

Reflection. What is it that you like about your favorite gaggle member? What is it that you dislike about your most annoying gaggle member? In what ways do these gaggle members interconnect with you? How close are they to you? In what ways do they interrupt you? In what ways do they augment (or displace) your work? What else do you notice about your use of these gaggle members? As you reflect on the gaggle what do you feel and where do you feel it?

Intentions, Intentionality, and the Gaggle

As the gaggle moves closer and becomes more interconnected with humans, as it interrupts us more frequently, and as it augments our work to the point of displacement, there grows a sense that somehow, we need to get a handle on these things—set boundaries perhaps—to preserve some of the good stuff of being human. For example, Sherry Turkle (2012) has warned us about the ways in which digital technology changes human relationships, and authors like Rob Smith (2019) have concerns that the gaggle is leading us to lives full of almost unavoidable bigotry.

To be sure, lots of good things come from the gaggle; if that weren't the case, it wouldn't be so darned attractive, seductive perhaps. Having said this, how can we approach the gaggle's gifts with greater intention? As we might with a nosey neighbor, how can we set boundaries on the ways we let the gaggle into our lives, so it serves us, instead of us serving the gaggle? Put somewhat differently, how can we enjoy the benefits of the gaggle and minimize the way it harms us as human beings?

These are important questions, and we believe that there is a linguistic clue to some good answers in the last paragraph in the word *intention*. To intend is to plan or want to do something and this is something that human beings are particularly good at. By contradistinction, computers, AI, and machine learning are, at present, decidedly non-intentional. If an AI plays chess, checkers, or Jeopardy, it has been programmed for that purpose. If it drives a car, translates a language, or does computer vision tasks, the intention is that of the programmer, not the program. If an AI invents something or creates music or art, again the inventive intention and the range of what can be invented is built in. Even sophisticated robots with the ability to roam around and make decisions based on where they are, have their range of goals or intentions decided ahead of time, and robots with emotionally expressive faces do not own their emotions; they are preordained by the programmer. In short, at this point in the evolution of artificial intelligence and machine learning, whatever intentions computers have, they are built into their programs by their programmers. This is an important distinction to keep in mind as it is a key to understanding what humans are good at, that computers are still struggling with.

This discussion of intentions leads to a larger discussion of what philosophers call *intentionality*—the ability to think about other things— and *intending* is only one part of the intentionality story. The richness of our thoughts, memories, feelings, and body feelings, and our ability to wrap those around the things that we think about is a special part of being alive, of being conscious, and to this point in the development of AI and machine learning no AI and no ML system is intentional in any serious way like a human being (or a bird or a squirrel, for that matter). This is not to deny the possibility of *artificial intentionality* (Goldberg, 2018), but understanding important characteristics of what computers can't currently do is helpful to rethinking education in a world where the digital gaggle is rising and increasingly competent. We shall return to intentionality and its connection to the methods of the manual in later chapters.

From Learn-then-Earn to Do-Learn Business Models

Even during the writing of WNE back in 2013 and 2014, one of the things we were wondering about is the ways in which innovation can take place— is taking place—is the fundamental business model of higher education.

Traditionally, students spend four or more years in on-campus programs, taking classes and paying tuition. We call this the *learn-then-earn model:* a student "buys" an education that consists of a set of courses; once they have completed those courses, they can then do meaningful things in the world. This way of thinking about education dominates our language: we talk about education as a product; students "shop" for classes, majors, and colleges; magazines give schools "best buy" ratings calculated by considering the return on investment from tuition. And universities' business models depend on the revenue stream from selling these courses. But while this idea—that one must first learn, in order to do—is pervasive, it is deeply inconsistent with an idea that permeates the book: often, *the best learning takes place when students are creating value*: doing real things that matter to real people in the world.

Of course, this is not a new concept. If you rewind the clock, apprenticeship was a dominant mode of education for much of history. And how does apprenticeship work? The apprentice works with the master craftsman to do real work for a third party—and it's still how we do graduate education.

In undergraduate engineering education, there are similar examples: modern real-world design experiences in the senior year go back to the 1960s (Dobrovolny, 1996); more and more schools are ensuring some sort of industrially sponsored experience before letting students loose on the world; and students consistently identify these experiences as among their most meaningful. There is also a long history of so-called cooperative (co-op) education in which students alternate learning on campus with extended internship experiences in real-world organizations.

And around the world, students themselves have recognized the limitations of classroom learning. For example, the so-called *Junior Enterprise* (*juniorenterprises.org*) organization helps students organize student consulting firms in a variety of professional disciplines, not just engineering, in which students perform paid projects for local companies. Junior Enterprise is particularly strong in Brazil (Brasil Junior, *brasiljunior. org.br*) and many large Brazilian companies give preference to university graduates with Brasil Junior experience.

But as powerful as these experiences are, the cost structure and learning model of higher education remains firmly in the learn-then-earn space. Experiences like co-ops and Junior Enterprise in general are at the periphery, optional, or "shoe-horned" into a traditional educational product that is built out of polished, packaged courses.

What might be possible, both for student learning and for the business model, if we questioned the learn-then-earn model? One inspiring example is Paul Quinn College (2021), an historically black institution in urban Dallas. Many of their students come from poor backgrounds, and often have struggled both with paying for college, and with finding a job afterwards. So, in 2015, recognizing how much these twin challenges of college finance and finding a job might help alleviate each other, they decided to become the first urban work college in the US to require student employment concomitant with student classwork. Said differently, every student at Paul Quinn holds a job, either on campus or at a local company. And to accommodate this, the college changed their class schedule, their curriculum, their graduation requirements, and they used the revenue from student work to drive tuition costs down. The jury's still out, but early results are promising. One first year student described the learning this way: At Paul Quinn, "you actually *get it* and you *get paid for it.*"

We believe that if we center undergraduate education—both the learning model and the business model—on the idea that learning and value creation can be mutually reinforcing, we might both improve higher education and address the financial challenges that dog it. Olin is starting to experiment with a model that—it is hoped—will generate real value for external folks, provide faculty with some developmental opportunities, and enable deeply meaningful educational experiences for students. It's an approach that bridges the theory-practice divide discussed earlier, and one that we think might benefit everyone.

Adoption of a Basis Set of Rigorous Soft Skills

In WNE we talked about the importance of executive coaching skills for the transformation of engineering education. Our view at the time was that the skills were central to helping teachers move from being the *sage on the stage* to the *guide* (coach) *on the side*, and we stand by that assessment. Yet, as we have continued to work in education reform and transformation of our assessment of the importance of these skills has been elevated, our thinking about what skills are crucial has crystallized.

In particular, the next chapter (chapter 4) addresses what have been called soft skills by first abandoning the term "soft" for the term "shift" and then discussing five categories of shift skill (and mindset shifts) in what we call *the five shifts*. The particularly important skill of curious listening and its constituent components NLQ = noticing, listening, and questioning is discussed in some detail in chapter 5. These skills are important for teachers, for students, and for dealing with the digital gaggle.

While the main part of our work on shift skills takes place in chapters 4 and 5, we do take some time in this chapter to talk about two shifts, in part because they are fundamental to how our thinking has grown around the challenges and crucial needs of educational change, and in part because it is hard to discuss our views on educational change without introducing them now. We start with co-contraries and continue with the problem of theory privilege.

Thinking with Co-Contraries

In 2014, we were aware of the concept of a co-contrary (or what some call a polarity), but it was not a central concept in the text. As we have worked to disseminate the WNE's ideas around the world, we have found co-contraries to be particularly helpful in promoting effective change, especially in overcoming the tendency of problem solvers to think of there being one right answer once and for all to complex problems. The idea of co-contraries has its roots in Chinese philosophy of yin and yang (Wang, 2021) as well as early Western philosophical ideas such as Aristotle's golden mean (Aristotle, 1992). Elbow has explored the notion of *contraries* as applied in writing (Elbow, 1983) and education (Elbow, 1987), and Barry Johnson has explored related ideas under the rubric of *polarity management* (2014, 2020). We take lessons from both authors and use the term "co-contraries" in all that follows.

A Gentle Introduction to Co-Contraries

As an adjective, the term contrary is to be "opposite in nature, direction, or meaning" and as a noun a contrary is "the opposite." For two contraries to be a co-contrary we recognize that in some important way or ways they are interdependent. In other words, a co-contrary is a pair of opposites or contraries that need each other; you can't have one without the other.

Inhaling && Exhaling as Co-Contraries

An illustrative example of interdependent contraries is given in the example of breathing: inhaling and exhaling (Johnson, 2014). Figure 3.1 shows the co-contrary *inhale && exhale*. We can ask, "Which is better, inhaling or exhaling?" and the question answers itself. We clearly need both. With a proper amount of inhaling, the organism gets sufficient oxygen; with too much inhaling, carbon dioxide builds up. With a proper amount of exhaling, carbon dioxide is cleared out; with too much exhaling not enough oxygen is supplied. In this way inhale and exhale are interdependent opposites or opposites that need each other. In this type of visualization, we list the benefits of a *sufficient* emphasis on each contrary in the corresponding upper quad and the difficulties that arise from an overemphasis on the corresponding contrary in the lower quads. In our lingo, these are the *sunshine* (good results from sufficient emphasis) and the *shadow* (deleterious effects from overemphasis) of the co-contrary, respectively.

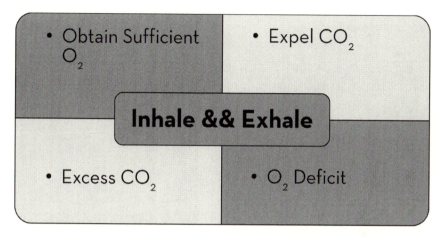

Figure 3.1: Breathing as the co-contrary *inhale && exhale* visualized in a sunshine-shadow quad.

From Breathing to More General Co-Contraries

The question we posed for breathing, "Which is better, inhaling or exhaling?" can be transformed to form a good rough-and-ready test of whether any pair of contraries A && B are co-contraries (we use the double ampersand as a connective symbol denoting co-contraries). If we ask the question, "Which is better, A or B?" and find that both are important, at least to some extent, then we are likely looking at co-contraries. In higher education we are sometimes concerned with several opposites: *teaching && research, teachers && students, theory && practice, freedom && structure, planning && experimentation, teamwork && individual work, competition && cooperation, diversity && homogeneity, unleashing && rigor* and so on. If you go pair by pair, and ask, "Which is better?" consistently we recognize the need for some combination of both teaching AND research, teachers AND students, planning AND experimentation, . . ., and unleashing AND rigor.

This last co-contrary (*unleashing && rigor*) recalls the discussion in chapter 2, where we presented the key findings of WNE. A primary result of WNE was to advocate a set of systemic changes that would *unleash* engineering students to have the courage to be innovative and creative: some have objected to this "solution" as they believed that it implied that engineers would no longer be *rigorous* in the application of their science, math, and engineering methods, but in light of this discussion, clearly *unleashing* && *rigor* form a co-contrary pair; WNE advocated for engineers who are both unleashed AND rigorous. Higher education is rife with co-contraries and just becoming aware of them is an important step toward making better change more effectively.

> **Reflexercise 3.2:** Make a list of co-contraries that are important in higher educational change. You may use the list of the previous section to get started (feel free to substitute your own labels). You may also consult the list of opposites on the web site (Polarity Partnership, n.d.) *assessmypolarities.com.*
>
> **Reflection.** Which co-contraries does your organization manage well, and which co-contraries are managed in an unbalanced? As you reflect on this question, what else do you notice about co-contrary management at your institution?

Acknowledging Co-Contraries Diminishes Fear of Change

In our work in higher education around the globe, we have found co-contraries to be especially important in educational change settings, because the fear of bad outcomes from change often causes hard resistance to any change at all.

The tendency to emphasize one contrary over another is sometimes called EITHER-OR thinking. We've observed that introduction of the notion of a co-contrary is helpful in that it can elevate concern for the ignored contrary. In our previous example of *unleashing* && *rigor*, someone on the rigor side of the debate can be presented with stories of accomplishment by someone unleashed and thus acknowledge that contrary's importance. Likewise, someone invested in unleashing can see the benefit of someone who approaches problem solving in a thorough and rigorous manner and be persuaded of the value of that contrary. In this way, the situation can move to an insistence on co-contrary excellence, toward a kind of AND thinking.

Stepwise, the move from EITHER-OR to AND can be achieved by (1) identifying key co-contraries, (2) reflecting on the good and bad results that can come from BOTH contraries (each contrary has benefits from proper emphasis and difficulties caused from overemphasis) and (3) assessing how the organization is doing right now on the particular co-contrary, and then (4) working to passively or actively manage, leverage, or move through the co-contrary to get good outcomes from both sides working together to achieve an excellent outcome more of the time. Of course, there may be many quite different solutions for managing any particular co-contrary, and there will usually be tradeoffs. In our running example of *rigor* && *unleashing*, the questions "How much rigor?" and "How much unleashing?" are always on our mind, but the act of calling out the co-contrary and explicitly valuing both sides often allows for changes that would otherwise be impossible without AND thinking; it can be somewhat unnatural at first, but once we embrace the power of AND (Johnson, 2020), it can become habitual in ways that lead to good approaches in complex settings.

Thus, by acknowledging the good effects of the currently dominant contrary, and by committing to achieving an excellent AND solution, not an EITHER/OR solution, educational change can be moved ahead more quickly because the parties' fear of losing the good aspects of the status quo is diminished.

Visualizing Co-Contraries with a Sunshine-Shadow Quads

The earlier visualization of the breathing co-contrary can be generalized to any co-contrary in the template of a *sunshine-shadow quad* in figure 3.2:

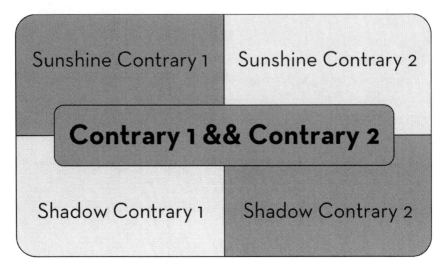

Figure 3.2: Template for a sunshine-shadow co-contrary quad.

The elements of a *sunshine-shadow quad* are as follows:

- Co-contrary label (label of contrary 1 separated from contrary 2 by the "&&" symbol).
- List of one or more bullet points describing good outcomes—sunshine outcomes—for sufficient or effective emphasis of contrary 1 (upper left) and contrary 2 (upper right).
- List of one or more bullet points describing deleterious outcomes—shadow outcomes—for overemphasis of contrary 1 (lower left) and contrary 2 (lower right).

With this understanding, let's fill in a blank sunshine-shadow quad for the co-contrary, *stability && change*.

Reflexercise 3.3: Print out a blank sunshine-shadow quad. Fill in the sunshine and shadow quads for the *stability && change* co-contrary per the text of the next section.

Reflection. As you fill in the quad, reflect on your own organization's stance toward *stability && change*. In what ways does your organization manage stability and change? What organizational structures, procedures, incentives, espoused values, artifacts, underlying assumptions, and personnel emphasize the importance of stability? What organizational structures, procedures, incentives, espoused values, artifacts, underlying assumptions, and personnel emphasize the importance of change? To what extent is one side the dominant contrary in your organization? What else do you notice about this co-contrary in your own organization?

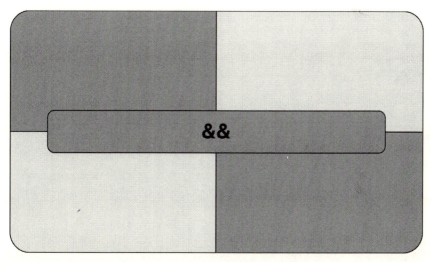

Figure 3.3: A blank sunshine-shadow co-contrary quad form.

Sunshine-Shadow Quad for *Stability && Change*

For example, with the co-contrary *stability && change*, we label contrary 1 (left of &&) "Stability" and contrary 2 "Change" (feel free to follow along by filling in a blank quad. If you prefer different choices of analogous language, substitute your own language).

In the left contrary sunshine quad (upper left) we might write several bullets: "Encourage continuity," "Preserve core values," and "Affirm systemic wisdom" (feel free to use your own language or add other virtues of appropriate stability). For contrary 2 sunshine results (upper right quad), we might write "Energize members and renew culture," and "Value little bets & innovation" and "promote continual learning (use your own language or add other virtues of appropriate change).

For shadow of "Stability" (lower left quad), we might write "Stagnate and resist adaptation," "Lose enthusiasm and engagement," and "Dismiss possibilities and opportunities." For the shadow of "Change" (lower right quad) we could write "Break working processes & systems," "Lose important values," "Take unnecessary risks." Look over the resulting quad chart and feel free to make changes important to you. When done you can look at a finished quad for the *stability* && *change* co-contrary below in figure 3.4.

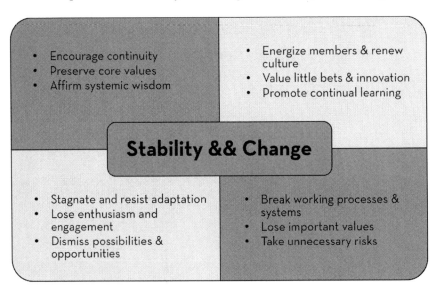

Figure 3.4: Completed *stability* && *change* sunshine-shadow quad.

The All-Too-Normal Dynamics of Contrary Lurch

One of the things that filling out a co-contrary quad helps us understand is the fundamental and quite normal dynamics of *contrary lurch*. For example, in our *stability* && *change* example, say we are in a good stable state (upper left quad), but we've been there for a while and our competitors are starting to get a foothold. At that point we start to lose position in the market, start to stagnate, and start to sense our loss of position. Recognizing this, we get the impulse to change, start to take some chances, feel the positive results from some of those, and start to the feel the sunshine of the change contrary (upper right), but in so doing, we will overdo it and lose some of the core values that made us a good organization in the first place. Thereafter, we notice what we lost from the good old days, try to hang on to our position from the new status quo without further risk. And so the cycle continues, lurching in a kind of butterfly pattern from the sunshine of one contrary to its shadow, to the sunshine of the next contrary to its shadow and so on and so forth.

How to Get the Best of Both Contraries: Stability Structures

Clearly, in the case of *stability* && *change*, we would like to get the positive effects of appropriate emphasis of both stability and change, and analogously for any other co-contrary, we would like to get the positive effects of appropriate emphasis of both contraries. But how do we do this in practice? It turns out, whether we recognize it or not, that we already manage co-contraries as individuals and in organizations using a variety of *co-contrary stability structures* or simply *stability structures*:

- Habits
- Time management and goal orientation
- Rituals & culture
- Organizational structure, control systems, and division of labor
- Active management and leadership

Here the term "co-contrary stability structures" refers to the ability to stably maintain excellence in both contraries of a given co-contrary.

As individuals, one key to managing personal and professional co-contraries is our adoption (and intentional modification) of *habits* (Duhigg, 2012; Clear, 2018). For example, in managing the co-contrary of *work* && *home*, the habit of getting up at a specified time, going to work, working a full day, and coming home at a pre-established time to spend time with loved ones is a common way to ensure you value commitments in two important parts of your life.

Similarly, using time management and a goal orientation can help us make sure that we do manage co-contrary activities that need attention. For example, faculty members may allot a fixed amount of time to course preparation to ensure that enough time is available for scholarly research (*research* && *teaching*).

At an organizational level, rituals and cultural norms help enforce default management positions on key co-contraries. A school that has a regular ritual of social events for students and faculty together is saying something about the importance of student-faculty relationships outside the classroom. A school that espouses the importance of balanced teaching, service, and research, but only promotes faculty members based on scholarly publication and funded research, should not be surprised to see that action speaking louder than words with teaching and service deemphasized, both before and after tenure and promotion decisions.

Oftentimes, the structure, roles, and control systems that an organization defines dictate attention to co-contraries. A school that has outside contracts go through the Dean's office will get a certain level of attention to contractual details, whereas a program that hires a contracts officer will get specialized attention to those details and more time for an academic officer to attend to school matters (*academics* && *operations*). A school that measures faculty research productivity and grantsmanship and values student evaluations as the major metric of instruction effectiveness will get attention to those measures.

Finally, managers and leaders are a kind of co-contrary stability structure: they play an important role in monitoring co-contraries and actively balancing them within their organizations. Through their plans, communication, and feedback activities, managers and leaders are constantly affecting how co-contraries are handled by the organization. Much of the literature places a good bit of emphasis on the vigilance of leaders, but our view is that co-contraries are less often actively managed by individuals and are more often managed by *systems*. Cultural, rituals, organization structure, systems, espoused values, and organizational language and norms are the stuff of co-contraries on the ground. Change these stability structures and an organization will respond by balancing co-contraries differently.

In the case of higher education, the very recognition of the importance of co-contraries can itself help the organization to understand the importance of difference in the organization. For example, recognizing research and teaching excellence as well as service and administrative excellence helps those who value one contrary understand the importance of the different work of others.

Reflexercise 3.4: Referring to exercise 3.3, print out a blank sunshine-shadow quad chart and fill in the blank template for an important co-contrary to your educational institution.

Reflection. Which contrary is currently dominant, and which is under-valued in your opinion? What shadow and sunshine outcomes are currently observable on the left and right terms? What else do you notice about this co-contrary in your organization?

In the next chapter, we examine co-contraries as one of the five shifts and its relation to the other four.

Challenging the Privilege of Theory over Practice

A key co-contrary in education is *theory && practice*. Going back to the beginnings of the modern university at the University of Bologna in 1088, we observe that the university was originally organized as an assembly of nascent experts in different areas of human knowledge (WNE Ch. 3, pp. 75-76). This tendency toward increasingly specialized knowledge and narrow expertise has deepened through World War II and into the current century to the point where universities offer education that largely assumes that practice is the mere application of well-vetted theory. MIT philosopher Donald Schön (1983) called this assumption *technical rationality* (TR) and suggested that it misunderstands the special nature of knowing in practice, what Schön called *reflection-in-action* (RIA). RIA is a dialectical process, often involving *reflection* and *conversation* among different practitioners, between practitioners and clients, and between practitioners and themselves. Here we call the tendency to think of practice as merely the application of theory a kind of *theory privilege*. In chapters 4 and 5 we call out a particular set of reflection and conversation skills that help us better understand and respect practice as something that deserves serious attention above and beyond the theoretical content of many higher education degree programs.

New York Yankees baseball player and manager Yogi Berra was known for his sticky paradoxical statements (e.g., "When you come to a fork in the road, take it.") and Yogi is often credited with the following statement:

In theory there is no difference between theory and practice.
In practice, there is.

A recent study says that Yogi didn't say this,[2] but whomever said it, we're with them. A major obstacle in improving education is the idea that practice is the mere application of well-vetted theory; understanding, teaching,

---

2. This quotation is often attributed to Yogi Berra. It is also attributed to Albert Einstein, Richard Feynman, David Jeske, and others. A nice analysis is presented at Quote Investigator (2018) that presents fairly compelling evidence that the remark was first made in February 1882 by Benjamin Brewster while he was a student at Yale. We will attribute the quote to Bishop Brewster and from time to time we may call it a *faux-Yogism*. Of course, Yogi covered situations like this with the authentic Yogism, "I really didn't say everything I said." (Berra, 2010).

and practicing the process of how we come to know things in practice more seriously is an important step toward the improvement of all education.

As noted above, theory and practice are a proper co-contrary, and we are not here suggesting that education should now lurch from theory to practice. Theoretical teaching and learning remain important parts of a student's education. What we are suggesting is that the royal road to education improvement involves managing the co-contrary between theory AND practice more thoughtfully and successfully. We will have more to say about this in the next chapter.

Substantive Educational Change in Months, Not Years

In 2014 we believed that substantive educational change in existing institutions was both difficult and required multiple years to see it through. These beliefs were challenged by a pilot project applying a new method for rapid substantive change. Chapter 6 considers this method, what we call *the four sprints and spirits method* or 4SSM. 4SSM accelerates substantive change in a principled way by embracing shift skills and two other action frameworks by engaging a carefully sequenced, deep set of exploratory and reflective conversations in a way that can result in both substantive change and a mood that embraces student learning and unleashing in the manner of *A Whole New Engineer*. Although we present this process as a rapid step-by-step approach to change, the elements of the four sprints and spirits may also be slowed down for those who wish to achieve their change more organically.

Affective Implementation Carries over the Mood of Change and the Plan

One of our more recent insights has been around the implementation of change plans. Processes such as 4SSM are quite good at coming up with a change plan, but all the deep listening, reflection, and conversation of 4SSM results in both a rational plan and a journey to a good mood. We now explicitly recognize that both products—the plan and the good mood—

are worthy of nourishment and implementation. We will take up this subject in chapter 7 as part of what we call *affective implementation*. One of the keys in this approach to implementation is to create a virtual or physical place to permit the sprouts of a new culture to take root in what we call a *respectful structured space for innovation* or RSSI. The processes and the place for affective implementation will be taken up in more detail in chapter 7.

> **Reflexercise 3.5:** Journal or make some notes about your takeaways and action-aways from this chapter.
>
> **Some questions**. Skim back over this chapter. What parts resonated with you? What parts did you disagree or have some concern with? What terms were used in ways that seemed familiar or unfamiliar? What experiences in your life connected to what you read? What questions arose in your mind for further reflection and consideration? What inspired you to action?

Dip or Dive

1. **Yogisms.** *Dip.* Go online and search for a listing of some of Yogi Berra's famous Yogisms. Choose 2-4 of them that seem relevant to educational change initiatives. *Dive.* Get or borrow a copy of *The Yogi Book* (Berra, 1999) and repeat the exercise.

2. **Inspirational education quotations.** *Dip.* Use *www.brainyquote.com*, *www.goodreads.com*, or some other quotation site and find 2-5 inspirational educational quotations that capture the spirit of your change initiative. *Dive.* Decompose the educational initiative into discrete elements or components and find 2-5 inspirational quotations for each element. Then, incorporate these into your pitch deck.

3. **Habits.** *Dip.* Watch the TEDx talk by Charles Duhigg (2013) *https://www.youtube.com/watch?v=OMbsGBlpP30* in which he describes key points from his text *The Power of Habit* (2012) or watch James Clear discuss ideas from *Atomic Habits* (Clear, 2018)

in this video (APB Speakers, 2018): *https://www.youtube.com/watch?v=U_nzqnXWvSo*. *Dive*. Acquire one or both of the books and journal about key insights.

4. **Business models canvases.** *Dip*. Watch Alexander Osterwalder explain the basic notions of a business model canvas (Osterwalder & Pigneur, 2010) in this video (Darus, 2015): *https://www.youtube.com/watch?v=RpFiL-1TVLw*. Apply the canvas to the business model of your current educational institution. *Dive*. Acquire the text mentioned above and repeat.

5. **Rituals.** *Dip*. Watch Kursat Ozenc discuss key concepts from the book *Rituals for Work* (Ozenc & Hagan, 2019) in the video (Stanford Alumni, 2019) *https://www.youtube.com/watch?v=dAroNh-P1UM*. In what ways can you bring rituals into the creation of an educational change initiative? *Dive*. Get the book and go deeper.

6. **Chinese room problem.** The problem of what AI can and can't do has been discussed using John Searle's famous *Chinese room argument* for over four decades (Cole, 2020). *Dip*. Watch the short video about the Chinese room here (Kaplan, 2020): *https://www.youtube.com/watch?v=tBE06SdgzwM*. What do you take away from the discussion? What portions are confusing to you? *Dive*. Read about the Chinese room in the *Stanford Internet Encyclopedia of Philosophy* (Cole, 2020) here: *https://plato.stanford.edu/cgi-bin/encyclopedia/archinfo.cgi?entry=chinese-room*. In what ways does the longer exposition clarify or confuse the issues for you?

Reflectionaire

1. **Teaching & learning.** Consider the time frame covered by this chapter (2014-present). What things have you learned about how education needs to be changed or can be changed? In what ways have your teaching/learning methods evolved in that time frame? In what ways has your educational institution changed hiring, tenure/promotion, and other advancement criteria to align (or misalign)

with superior teaching? What else do you notice about teaching and learning in that time period?

2. **Research.** Consider the time frame covered by this chapter (since 2014). What things have changed regarding research in that period? In what ways are publication, funding, and other parts of the research game more or less competitive? In what ways has your approach to research and/or leading a research team evolved in that time frame? To what extent is it easier or harder to find quality graduate students (professors)? What else do you notice about research now versus earlier? In what ways have your educational institution's criteria for research in hiring, tenure/promotion, and other advancement changed?

3. **Theory privilege & the misunderstanding of practice.** Consider the way your discipline is traditionally taught. In what ways is theoretical knowing elevated above knowing in practice? In what ways is the tendency counterbalanced by a deeper understanding of practical knowing, and what kind of experiences, coursework, materials, and teaching help students come to practice as a distinct or different way of knowing? What else do you notice when you reflect on theory and practice?

4. **Key co-contraries for your change initiative.** What are the key co-contraries that must be managed, navigated, or leveraged in your change initiative? What are the current preferred contraries? In what ways are substandard results being obtained because of under-developed contraries? In what ways can you move from EITHER/OR thinking to AND thinking in your initiative? What key things from the status quo are at risk from excessive emphasis on change? What are the key risks of not changing? What else do you notice about these co-contraries as you reflect on them?

5. **DIY (do it yourself).** Upon reading this chapter, what concept, experience, innovation, or other object caught your attention that deserves further reflection? What is it about this object of your attention that interests you? What have you experienced in

connection with this object? What other questions need to be asked and reflected upon in connection with your DIY reflection-aire? Reflect on those questions. What else do you notice as you reflect upon this object?

Little Bets

1. **Co-contrary quadding party.** Referring to reflectionaire 4 above, pair up or form small groups and make extended lists of relevant co-contraries in professional and personal life, then vote for your favorite or most relevant ones. Form small groups around the top co-contraries of interest, and then create a sunshine-shadow quad for each one. What little bets are ripe for action based on these analyses?

2. *Reflective Practitioner* **book club.** To dive more into the nature of knowing in practice as distinct from knowing in theory, gather a group of interested people and read *The Reflective Practitioner* (Schön, 1983). Consider the following questions: In what ways does the book resonate or conflict with your thoughts and feelings? To become a more reflective and conversational practitioner, what skills can you develop in line with Schön's stories of practice. What conversations are important to have right now in the context of education and how can you make some little bets in initiating them?

3. **Business models of education.** Gather a small group to consider the business models of education. Historically, there are other models of education besides the completion of some degree at a school, things such as cooperative education and apprenticeships. With the advent of the internet, there are many attempts to modify, augment, or supplant current models with technology. Using web search freely and loosely, make lists of all the different business models of education you can find. Use the book *Business Model Generation* (Osterwalder & Pigneur, 2010) to consider the ways in which these models function. Consider hybrid combinations of current models that may be more appropriate for our time and technology.

4. **DIY little bet.** What little bet needs to be made as a result of your thinking about the material in the book so far? What is easiest to try or do with existing means? What is the smallest possible experiment that will get you some interesting results? What do you notice from doing this experiment? What else needs to be done?

4

5 Shifts for the Cultivation of Connection & Insight

From "Soft" Skills to Shift Skills

The notion of reflection-in-action or RIA briefly discussed in the previous chapter in connection with the problem of theory privilege is critically important to the improvement of education, and in WNE (Ch. 9), we talked about the importance of executive coaching skills in both the educational change process and in the curriculum itself. We stand by the importance of both emphases, but the term "reflection-in-action" is not widely known, and the term "coaching" can be problematic on ego grounds because the use of the term is a tacit admission that someone needs help from another.

To make matters worse, these important skills are often lumped under the rubric of what others like to call *soft skills*, and the term "soft" can have connotations that such skills lack rigor, are not particularly important, or that they are easy to master. Furthermore, when people talk about soft skills, they often bring forward a very long list: presenting, speaking, sales, conflict management, persuasion, negotiation, teamwork, collaboration, entrepreneurship, planning, and so on and so forth. Each of these skills can be important in practice, but the long list of disparate skills is part of the problem because it doesn't get to the core or the essence of the skills that everyone needs in everyday life. In figure 4.1, we depict some of the long list of what we call *derivative soft skills* on the outer ring, and the visualization begs us to question how these ostensibly different things relate to each other, and what core skills are shared and basic to those on the list.

Shift Skills and the Cultivation of Connection & Insight

To move toward better educational practice and reform, here we do three things. First, we take some care in the labeling of the skills that we believe are the most important. To do so, we drop the term *coaching skills* on ego grounds; we also stop calling them *soft skills*, because we believe that they are often as important as disciplinary skill, and that they can and should be approached explicitly, seriously, and rigorously. Having decided what not to call them, we should also take some more care in what we do call them, and this leads us to three words.

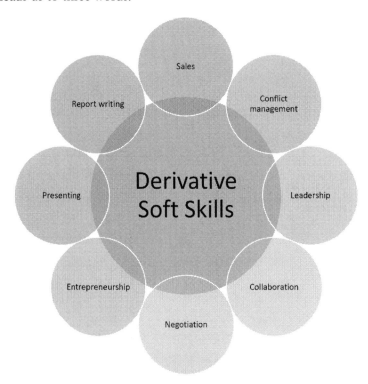

Figure 4.1: Derivative soft skills are many, and their interrelationship is unclear.

The first word we use is the term *shift*. We call these important skills *shift skills* because they help us bring about great personal and organizational change—they help us *shift*. These skills, whatever we call them, are as important—sometimes more important—to life success as great disciplinary skills. Moreover, we live in very fast-paced time in which technology and society are changing rapidly and unpredictably. Great shift skills help us keep up with this relentless pace of change.

The second term we use is *connection*. In reviewing WNE, we discussed the components of unleashing and one of them was *connectedness, collaboration, and community*. We affirm the importance of this unleashing component, and here we examine the role of one-on-one connection in ways that direct the five shifts toward better understanding of others and greater connection with them.

The third word we use is the term *insight*. Shift skills help us in making change, and they do so through a particular mechanism; they help us have a series of insights that improve our ability to take effective action. We will say more about mechanisms of insight later in this chapter.

Skills and Mindsets

Second, we recognize that when we modify or augment a skill, there is often a reframing of mindset that accompanies it. For example, we have already discussed the fundamental mindset shift of WNE as moving from thinking about content, curriculum, and pedagogy as primary change variables to thinking of emotion and culture as primary. In this book, we will use the term *shift* to capture both *mindset shifts* and the shift *skills* we use in practice.

Five Core Shifts

The third thing we do is recognize that not all shift skills are equally fundamental. We now believe that the derivative soft skills can be thought of as being composed of a combination of what a mathematician might call a five-dimensional *basis set*—the *core shifts*. In mathematics, a two-dimensional space requires two independent coordinates to locate a point in the plane, a 3-D space, three independent coordinates in the cube, and so on. Analogously, we believe that derivative soft skills are largely made up of a combination of five shift skills.

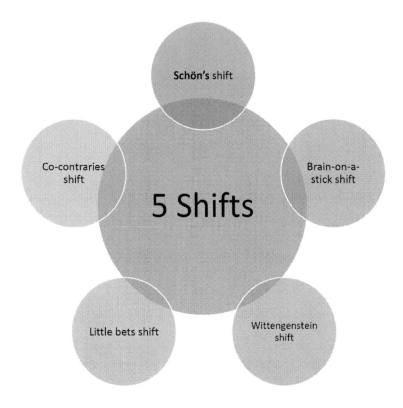

Figure 4.2: The five core shifts form a kind of basis set for derivative soft skills. In other words, derivative soft skills can be expressed as some combination of the core skills.

Elsewhere, one of us (Goldberg, 2018, 2019b) has listed and labeled the five dimensions of shift skills—the five core shifts or simply the five shifts—as follows:

- **Schön's shift**: The shift from practice as the application of theory to particular situations to practice as reflection- and conversation-in-action.

- **Brain-on-a-stick shift:** The shift from head to heart, body, and hands.
- **Wittgenstein's shift:** The shift from language as description to language as action.
- **Little bet's shift:** The shift from planned action to entrepreneurial thought and action.
- **Co-contraries shift:** The shift from problem solving to managing or navigating co-contraries.

In what follows, we discuss each in turn. In addition to the in-line reflexercises of previous chapters, in this chapter, we introduce a new type of box, the *experience box* or simply *E-Box* in which we share actual or somewhat fictionalized experiences (drawn from a composite or realistic extension of our experience) of chapter material.

Schön's Shift: Combining Knowing in Theory AND Practice

Earlier we poked fun at theory privilege with the faux-Yogism, "In theory there is no difference between theory and practice; in practice there is." This bit of word play helps us think more deeply about the ways in which practice goes beyond the mere application of well-vetted theory. MIT's Don Schön called the idea that theory is merely the application of theory to particular situations *technical rationality* and he distinguished it from the idea that practice is often a deeply reflective practice using the term *reflection-in-action* (RIA); many of the examples of RIA in his book *The Reflective Practitioner* (Schön, 1983) are *conversational* in nature.

Schön emphasized the importance of practice as a kind of reflection about the situation of practice and possible actions, and we couldn't agree more, but in a busy world like the one we now inhabit, taking time to reflect individually feels like a luxury that many of us cannot afford. For this reason, and because it is easier to learn to help others reflect and then apply that learning to our own individual reflection processes, herein we emphasize the importance of reflective conversations as the royal road to effective reflection- and conversation-in-action, both.

Specifically, we emphasize conversations driven by *curious listening*, and we use the acronym NLQ to stand for noticing, listening, and questioning as the key components of curious listening that help others reflect. In the next chapter, we repurpose the teaching sequences and exercises used routinely in training executive coaches to help give educators and students the key skills of curious listening. In workshops and classes over the last decade all over the globe, we have had considerable success in bringing these skills to both faculty and students to the point where we strongly believe that more widespread adoption of these methods is in order. In the next chapter, we return to the practice of curious listening and NLQ skills with a series of exercises.

Brain-on-a-Stick Shift: Combining Head, Heart, AND Body

The brain-on-a-stick shift refers to a famous PhD comic (Cham, 2017). In the first panel a graduate student thinks of herself as complex, full of hopes, dreams, and aspirations. In the second panel, a professor's view of the student shows a picture of the graduate student as a brain-on-a-stick, and the professor off panel asks, "So how's research?"

Figure 4.3: Brain on a stick comic. "Piled Higher and Deeper" by Jorge Cham *www.phdcomics.com*. Reprinted by permission.

Engineering and STEM education is organized to fill brains-on-sticks with technical knowledge, and increasingly humanities, social sciences, and other professions are organized around particular theoretical concepts. A key shift in higher education is to move from thinking of students (and humans more generally) as brains-on-a-stick and to think of them as embodied individuals with thoughts, feelings, and body sensations. There are various practical approaches to so doing, and we believe that this shift is particularly important to practitioners who develop solutions that serve human beings more fully and to organizations and institutions that are as caring and humane as they are efficient and productive.

Wittgenstein's Shift: Viewing Language as Description AND Creation

Ludwig Wittgenstein was a philosopher in the last century who was concerned with how language works. His early contributions took the position that language is a kind of formal, static description of what is. His later contributions took a different view—the opposite view—and he argued for language as a kind of generative action. This latter viewpoint has led to an understanding of language in *speech acts* theory (Searle, 1999), where phrases with the same propositional content can have different functions or force ("you are sitting" versus "please sit"). Sometimes language is committed to *asserting* objective facts, to offering *assessments* or interpretations, to making a *request* (or giving an order), to making a promise or a *commitment*, or to simply making a *declaration*. One of the intellectual foundations of modern executive coaching is to understand speech acts and how they can affect teamwork, trust, and the smooth function of organizations more generally (Flores, 1981). We believe that instilling an understanding and application of speech acts in faculty and students is an important step forward for higher education.

Little Bets Shift: Combining Planning AND Effectuation (Experimentation)

The reliance of higher and professional education on the idea that practice is the application of extant theory results in a pervasive overreliance on (and sometimes overconfidence in) the efficacy of *planning*. The success of planning, by its very nature, depends on successfully predicting a sequence of future outcomes with relative certainty and reasonable accuracy. This works well when we have effective theory for or a good deal experience with—preferably both—all the events being planned in the sequence, but it can fail miserably in relatively un- or underexplored domains with incomplete theoretical foundations or little experience.

Of course, these are environments where entrepreneurs rush in, and Sarasvathy's groundbreaking work (Sarasvathy, 2008) on entrepreneurial thought and action distinguishes between routine planned work and the entrepreneurial kind on exactly these grounds. Sarasvathy coins the term *effectuation* to remind us that entrepreneurs make plans and expect them to fail in part because of the uncertainties involved. After initial action taking, the entrepreneur takes subsequent action starting from what has actually happened—from the *effects* of previous action—not from what the entrepreneur hoped would happen. In this way, the entrepreneur iteratively is led to something useful, something that may be quite different from the original goal, but something useful, nonetheless. By contrast, someone with a planning mindset, will often stick with the plan and try to make it work, sometimes beyond the point of diminishing returns.

It should be clear from the context that we are using the term "entrepreneur" in a way that is almost entirely disconnected from profit-making and almost entirely connected toward a person's attitude toward experimentation in relatively unexplored domains. In the sense here, someone can be an entrepreneurial engineer, or physicist, or social scientist, or humanist, or party planner, or hair colorist, if one goes off road and explores unknown paths with little experiments, thereafter, using that knowledge as the starting point of subsequent exploration.

Reflexercise 4.1: Journal or make some notes about the 5 core shifts.

Some questions. Which of the 5 shifts is of most interest to you? In what ways do you already practice the 5 shifts, and in what ways would you like to deepen your practice of them? Which of the 5 shifts gives you the most difficulty at work? What other thoughts and emotional and body feelings come up for you as you reflect on the 5 shifts? What else do you notice?

Co-Contraries Shift: Managing Co-Contraries AND Problem Solving

The final shift—to co-contraries thinking—has already been discussed at some length in chapter 3. Recall that we recognized co-contraries as pairs of opposites that need each other, such as *inhaling && exhaling, direction && freedom, teaching && research, individual work && teamwork,* and *unleashing && rigor.*

Here we formally recognize co-contrary thinking as one of the core shifts. Balancing EITHER-OR thinking and AND thinking is as important as the other four. Moreover, we notice something quite interesting. Each of the five shifts discussed in this chapter is itself in co-contrary with a core value of the higher education status quo. Practice is dominated by theory. Heart and hand are dominated by head. Language as action is dominated by language as description. Experimentation is dominated by planning. Finally, AND thinking and co-contrary management are dominated by EITHER-OR thinking and problem solving.

Understanding this point is crucial. We declare the importance of these five co-contraries at the same time we recognize the dominance of the five status-quo contraries over each of the five shifts. In other words, as we move each of the shifts out of the shadow of its dominant contrary, we explicitly recognize the need to manage it with respect to its opposite; we seek AND excellence on each of the five co-contraries. Said differently, when we argue for the shifts, we are not arguing for EITHER/OR thinking. We are arguing for managing each of the five co-contraries well for more effective educational institutions educating more effective graduates in a world that values key differences in thought and action with relish, not resistance.

The Five Shifts and the Digital Gaggle: A Human-Gaggle Comparison

We started the book by suggesting that much of what we were talking about was relevant to recrafting education in a world in which internet, smartphone apps, distributed access to digital information, artificial intelligence, machine learning, and robotics were rising. This *digital gaggle* is and will continue to be quite disruptive to human beings and their work, and the gaggle is becoming increasingly competent in certain narrow kinds of ways. Having said this, if we review the five shifts, we see that each of the shifts represents something that humans can do quite well and concomitantly is something the digital gaggle does quite poorly or not all. Let's look at this shift by shift via a human-gaggle comparisons or HG comparison:

- **Schön's shift**: The shift from practice as the application of theory to particular situations to practice as reflection- and conversation-in-action. **HG comparison:** The digital gaggle can already apply theory in narrow and well-defined areas of human endeavor and will continue to get better at so doing, but the gaggle is not good at general reflection- or conversation-in-action on broad topics of interest to human beings in any real sense.
- **Brain-on-a-stick shift**: The shift from head to heart, body, and hands. **HG comparison:** The gaggle is great at certain narrow rational tasks, and modern robotics is increasingly capable of certain well specified physical functions and manipulations; however, the gaggle has no equivalent of true emotion (smiley face robots simulate emotion but have no real emotion) and robots do not get gut feelings about what to do in emergency situations (Klein, 1998). Human beings can be great at head, hands, and heart and their integration.

70

- **Wittgenstein's shift:** The shift from language as description to language as action. **HG comparison.** Natural language processing on computers can be said to "understand" language in some narrow and formal sense that matches the idea of language as description, but computers have no intentionality (the ability to think about other things) and the only actions they can take are those their AI/ML circuitry is permitted to do. There is generally no sense in the digital gaggle of using language to create something new, while humans can do this almost casually.

- **Little bet's shift:** The shift from planning to entrepreneurial thought and action. **HG comparison.** Early AI systems were well concerned with planning (Feigenbaum & Feldman, 1963), and this early trend has continued; planning assumes good causal knowledge from theory or empirical experience where one can work backwards from a desired state to the current one. Having said this, the gaggle doesn't take entrepreneurial flyers on experiments that might lead to something interesting, whereas humans can be quite good at this.

- **Co-contrary shift:** The shift from problem solving to managing co-contraries. **HG comparison.** The gaggle is oriented to narrow problem solving in a big way and can be quite good at it, but the ability to do meta-level jumps to seemingly unrelated opposites is beyond their ability, now and into the future. The human balancing of sophisticated difference is something that humans can learn to do quite well.

Reflexercise 4.2: In what ways has the digital gaggle (the web, smartphone Apps, artificial intelligence, machine learning, and robotics) impacted your life.

Some questions. Which impacts do you view as positive? Which as negative? What areas of your work and home life are sufficiently narrow or rote, where the gaggle may soon be as competent as humans? In what ways are you gaggle-proofing your work skills to stay relevant? What else do you notice about the impact of the digital gaggle?

E-Box: From robot to human. About 5 or 6 years ago, Mark and Dave were training young management consultants at a major firm in South America, and we could tell that the material of this chapter was really resonating with many of them. We didn't know how much until one of the members of that cohort contacted us recently and told stories of personal and professional change.

For example, the modules on *planning && experimentation* led to one of our contact's colleagues to move on and become quite successful in start-up entrepreneurship rather than management consulting.

Our contact reported how the material changed his life in the sense that learning about curious listening, NLQ, and the brain-on-a-stick shift helped change this engineering trained individual from "a robot to a human." Almost immediately he was empowered to move on to a successful career in non-profit leadership and social entrepreneurship, creating educational opportunities for the under-privileged in his home country.

Lesson. The 5 shifts are sneaky impactful. They sound obvious, but with a modicum of practical application they amplify personal and professional effectiveness.

As we can see, the five shifts line up as a set of pivotal things that human can do well, and the gaggle has trouble with. Given that the current arc of the digital gaggle is to take over more and more narrow tasks quite competently, it behooves educators to spend more and more time on exactly those areas where human effort will continue to be distinctive—those areas where humans have and will continue to have a natural competitive advantage. The five shifts are quite different, but when we set them side by side—when we think about reflection, conversation, and practice, integrating head, heart, and hand, using language descriptively and as action, planning and experimentation, and managing co-contraries and problem solving, a term that seems to capture the net result of these activities is *insight*. "Insight" is one of those slippery terms that can be used in different contexts, so let's see if we can be somewhat more precise.

What is Insight?

For our purposes, insight may be defined as follows:

> *Insight is a change in how we interpret what we sense in a situation that enables different sets of thoughts, feelings, and actions that in turn help us obtain results that serve us better*

The traditional coaching model used for understanding insight is the so-called OAR model (Brothers, 2005), which originates in the work of Argyris and Schön (1974, 1978).

O-A-R: Observer-Action-Results

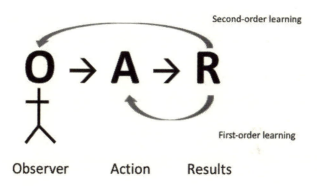

Second-order learning

First-order learning

Observer   Action   Results

Figure 4.4: O-A-R model represents the observer taking action, getting results, and subsequent learning either by changing the action or changing the interpretive complex.

In the model, an observer O takes in sensory inputs (including natural sensory data, language, and other symbolic inputs) and *interprets* them in particular ways that permit actions A to be taken that winds up in results R. The observer's *interpretive complex* can be quite involved, encompassing such things as values, beliefs, goals, models, distinctions, past experiences, stories, or other structures in the mind. The *action complex* can also be quite involved permitting

73

changes in physical, cognitive, and emotional state internally, and physical or speech actions externally. The results of this chain of events have both physical, cognitive, symbolic, and emotional consequences, both internally and externally.

As humans we are constantly learning from our actions, and a normal mode of learning would be to try different actions under relatively constant interpretation. This kind of learning is called *first-order learning* as shown in the figure as indicated by the arrow going from the result back to action. In other words, we modify the choice of particular actions (under relatively constant interpretation) to improve results. This is quite normal, and first-order learning is a common form of learned improvement. Having said this, sometimes, no matter what action we take, we get substandard outcomes, and this non-improvement signal begs us to see if we can interpret our inputs in a more productive manner.

In these non-improvement cases, we are led to somehow (more on this in a moment) change the observer O's interpretative complex (we could call this a move from O →O'), which in turn may open a different set of possible actions (from A→A'), call them A'. Now, when a similar situation comes up, the observer O' interprets the inputs differently, which permits different actions to be taken with the consequence that results obtained are much improved. This kind of learning is called *second-order learning* as indicated in the figure by the arrow going from the result back to the observer's interpretation scheme. In other words, we modify observer interpretation schemes to possibly open up different action sets to thereby achieve better results. This kind of human learning is also quite normal, but it is a bigger deal than the stimulus-response kind of learning of the first-order scheme. The achievement of better results through improved interpretation is what we call "insight."

Three Ways of Having Insight

The foregoing discussion helps us understand the importance of modifying our interpretation scheme to the achievement of insight. Let's consider three common modes of interpretation improvement:

    1.   Injected distinctions & models

2. Elevated awareness and modification of current distinctions
3. Reframing of key distinctions

Much of how we interpret the world is carried in our *distinctions;* when we make a distinction, we use a term or terms to discriminate between things as different and in this way, by making finer and finer distinctions we are able to see things differently in ways that permit useful action.

Thus, one of the ways we gain insight is by simply *injecting* already understood distinctions and models. In the next chapter, we will explore a number of distinctions around noticing, listening, and questioning that will make deeper listening easier. In this way, those distinctions are *injected.* Much of traditional higher education is the injection of well-worn, but very useful practical and theoretical distinctions.

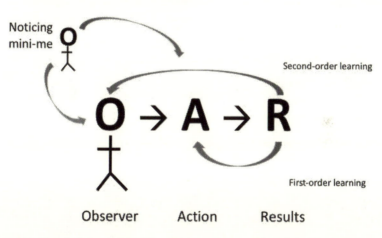

O-A-R + Noticing

Figure 4.5: The O A R model with noticing mini-me. The ability to notice our thoughts and actions allows for a kind of meta-level of learning.

Another way we make interpretive changes that lead to insight is through *elevated awareness and modification of current distinctions.* This can be seen in the modified OAR model with the noticing mini-me. In this figure, we see the core OAR model, and the noticing mini-me off to the left.

We will spend more time in the next chapter on noticing and awareness, but for now, suffice it to say that there is a part of the brain (in the prefrontal cortex) that can be aware of what other parts of the brain are thinking or feeling. When we exercise that part of the brain and have awareness of what we are thinking or feeling, we can stand back from those thoughts and feelings and sense that they may be contributing to our inability to get good results. That awareness allows us to modify our distinctions and doing so permits different interpretations, possibly different actions, and possibly better results.

A third way we modify the interpretive apparatus productively is through reframing our distinctions. This may be viewed as a stronger version of item 2. In the reframing of distinctions, we tell a different story that is designed to evoke different thoughts and feelings that pave the way for better actions and results. These ideas connect with modern notions of *explanatory style* in positive psychology (Seligman, 1991) and connect to related notions in coaching (Caillet, 2008).

Reflexercise 4.3: Think of a recent time when you had an insight or a series of insights. These don't need to be earth-shattering or Nobel-prize winning, just simple insights that helped clarify some work or home issue.

Some questions. Which of the 5 core shifts was operational in your insight? In what ways did your interpretive complex (goals, values, beliefs, models, distinctions, stories, or other structures) change as a result of your insight? To what extent were any of the 3 mechanisms of insight at work? What else do you notice as you reflect on your insight?

Five Shifts as a Quadruple Whammy

Seen in this way, the five shifts are important in four ways. First, they help faculty become more coach-like and thereby more effective in unleashing students as recommend in WNE (Ch. 9). Second, they help faculty and other education change agents become more effective in leading and participating in change initiatives. Third, they help us AI-proof our work against the continuing onslaught of the digital gaggle. Finally, because of how the five shifts work together, they help us learn to facilitate human insight in others and then ourselves.

In the next chapter, we take up key conversational skills that will help us listen our way into facilitating human insight.

Reflexercise 4.4: Journal or make some notes about your takeaways and action-aways from this chapter.

Some questions. Skim back over this chapter. What parts resonated with you? What parts did you disagree or have some concern with? What terms were used in ways that seemed familiar or unfamiliar? What experiences in your life connected to what you read? What questions arose in your mind for further reflection and consideration? What inspired you to action?

Dip or Dive

1. *The Reflective Practitioner. Dip.* Watch the video or skim through the transcript of Donald Schön's Iowa State talk (Mosior, 2021) based on his 1983 text, *The Reflective Practitioner: https:// hiredthought.com/2021/02/24/donald-a-schon-at-iowa-state-university-talk-transcript/.* What do you take away from this talk for the reform of higher education in your institution and generally? *Dive.* Read the book and answer the same questions.

2. *Your Brain at Work. Dip.* Watch David Rock's Georgia TechTalks (2009) presentation (*https://www.youtube.com/ watch?v=XeJSXfXep4M*) about his book (Rock, 2009), *Your Brain at Work.* In what ways do these neuroscientific connections tie to the five shifts? *Dive.* Get and scan or read the text.

3. **Effectuation or little bets.** *Dip.* The ideas of being effectual (Sarasvathy, 2008) or making little bets (Sims, 2011) are explored in the *Big Beacon Radio* podcasts *https://www.voiceamerica.com/episode/92149/effectual-entrepreneurship-an-interview-with-sarasd-sarasvathy* (Goldberg 2016) and *https://www.voiceamerica.com/episode/95621/little-bets-and-breakthrough-ideas-an-interviewwith-peter-sims* (Goldberg, 2016). Listen to one of them and take notes on key ideas and insights. What key ideas do you take away from your listening? *Dive.* Get one of the books and skim, scan, or read it. Journal as you do this and contrast these ideas with the normal notion of planning.

4. ***Language and the Pursuit of Happiness.*** *Dip.* Listen to the Big Beacon radio podcast with Chalmers Brothers talking about the concepts in his book *Language and the Pursuit of Happiness* (Brothers, 2005): *https://www.voiceamerica.com/episode/96378/language-and-the-pursuit-of-happiness-an-interview-with-chalmers-brothers* (Goldberg, 2016). What parts of the conversation are salient for educational reform? *Dive.* Read the book and reflect on its lessons for educational reform.

Reflectionaire

1. **Schön's shift.** Consider the faux-Yogism: "In theory there is no difference between theory and practice. In practice, there is." Think of a time when you've experienced this sense of theory departing from reality. Consider these instances. In what ways was theory different? What steps were taken to overcome theoretical shortcomings? What else do you notice about these experiences?

2. **Big boys and girls don't cry (or feel).** Consider a time where strong emotions or body feelings were pushing you to do something or not do something, but your rational brain overrode those feelings. What was the overall result of the overriding? To what extent were your body feelings or emotions correct or incorrect? What else do

you notice about the experience? What can you learn from this and other similar experiences?

3. **Language in action.** Think of a time where you used the generative power of language to create a new class, program, or other initiative. What linguistic constructs were helpful to you in this endeavor? To what extent, did participants and others start to use your language? What else did you notice about using language in this creative way?

4. **Planning and little bets.** Sometimes organizational action is a combination of planning and little bets. Think of an educational initiative, new course, new program, or other change project. To what extent were planned elements executed as planned? To what extent were planned elements treated as little bets and modified because the original plans did not pan out? What else did you notice about organizational action in this setting?

5. **Key co-contraries in your life.** What are the key co-contraries that must be managed, navigated, or leveraged in your personal or professional life? What are the current preferred contraries? In what ways are substandard results being obtained because of underdeveloped contraries? In what ways can you move from EITHER/ OR thinking to AND thinking in your initiative? What key things from the status quo are at risk from excessive emphasis on change? What key opportunities are being missed because of excessive emphasis on the status quo? What else do you notice about these co-contraries as you reflect on them?

Little Bets

1. **Five shifts little bet.** This chapter has covered a fair amount of territory. Which of the five shifts seems most actionable in your personal or professional life, or in a change initiative you are associated with? What small action could you take that would advance your plans and goals? What would it mean to *smallify* that action somewhat? What would it mean to hypertrophy that little bet somewhat? What existing network, resources, and other means do

you have at your disposal that would allow for action sooner rather than later? What is holding you back?

2. **DIY little bet.** What little bet needs to be made as a result of your thinking about the material in the book so far? What is easiest to try or do with existing means? What is the smallest possible experiment that will get you some interesting results? What do you notice from doing this experiment? What else needs to be done?

5

Curious Listening: Keystone Habit of Educational Change

Charles Duhigg (2012) recounts the story of former Alcoa CEO Paul O'Neil using *safety* as a *keystone habit* to resuscitate a company with any number of financial and operational problems. Traditional managers thought this approach was strange and recommended more drastic and direct measures, but by focusing relentlessly on safety, O'Neill reasoned that many of the company's financial and operational shortcomings would be fixed appropriately, and he was right. Under O'Neil's safety program, Alcoa went on to raise earnings, increase share prices in a matter of a few short years.

Similarly, we can ask, is there a keystone habit for resuscitating flagging educational programs or for creating new ones? In our experience, the single greatest boost for robust education is *curious listening,* which we cover using the acronym NLQ or *noticing, listening, and questioning.* In the same way that business experts were skeptical of O'Neill's emphasis on safety, educators and educational stakeholders may be skeptical of the power of curious listening, but a crucial failure of universities as currently conceived is that they are top-down knowledge transmission systems in which administrators and professors are largely output devices delivering expert messages to students conceived largely as mere input vessels. As discussed in the last chapter, the five shifts are exactly those skills and mindsets that permit humans to thrive in the midst of the rise of what we called the digital gaggle—the aggregate effect of digital technology, artificial intelligence, machine learning, and robotics that is taking over more and more human work. And the central skill needed to call forward the five shifts is curious listening. Moreover, the central skill to bringing about change in higher education to educate humans to thrive amidst the gaggle is curious listening. Much of the misalignment, discontent, and dysfunction of modern universities would not exist if the practice of education were more of a curious listening enterprise in which stakeholders from top to bottom—board members, administrators, faculty, staff, students, alumni, and employers—all listened to one another.

In this chapter, we examine the three main components of curious listening, starting with the unexpected power of noticing.

Noticing

> **Reflexercise 5.1:** As you are reading these words, right here, right now, what do you notice? There is no right or wrong answer. You notice what you notice. Make a few notes on a piece of paper or in your journal about what you notice, and as you go through this section, reflect on your noticing and the section material, both.

On the first day of coach training at Georgetown University, one of us walked into class with 35 other student coaches and was immediately asked, "What do you notice?" This was followed swiftly with a strong negative emotional reaction, "What do you mean, what do I notice? I just paid five figures to be trained as a coach, so let's get on with it and give me some real coaching skills!" Of course, this reaction was as premature as it was wrong, and arguably the biggest part of a coach's job is high-quality noticing of what the client is aware of (or not) to help them have insights so they can overcome those things that are holding them back or take advantage of opportunities they may be missing.

The centrality of noticing to change is captured in a quotation by psychologist Daniel Goleman (1985, p. 24) in the form of one R. D. Laing's knots (1970):

> *The range of what we think and do is limited by what we fail to notice. And because we fail to notice that we fail to notice, there is little we can do to change; until we notice how failing to notice shapes our thoughts and deeds.*

Although Goleman is engaging in a bit of Laingian word play here, his point about the centrality of noticing to change is very important and seems obvious after it is stated; noticing is something that is easy for us to overlook or take for granted. To what extent are we really aware day to day, or to what extent are we creatures of our finely honed habits? How often do we jump to conclusions or make assumptions? What biases do we have for the things that we do and do not notice? To what extent can we move in and out of heightened awareness when it serves us? These are important questions, and it is largely around the ability to notice that executive coaching has become a global phenomenon. Coaches help clients at first by helping them become aware of thoughts, feelings, and body feelings they may not be aware of. Change thereafter isn't necessarily easy or simple, but with increased noticing and awareness, it is, at least, possible.

What Do We Notice?

Noticing is the starting point of change, but what can we notice? One distinction to make is that we may notice language, emotions, and body sensations as depicted in the figure below. For adults who have gone through any length of traditional schooling, language is a primary domain of noticing. We notice our thoughts in language, our self-talk in language, and the language of others, and given language's importance in human life, this emphasis on language is not unusual or weird. Having said that, noticing of emotions and body sensations and feelings can be particularly productive and helpful in many situations, and language noticing can sometimes crowd out these other kinds.

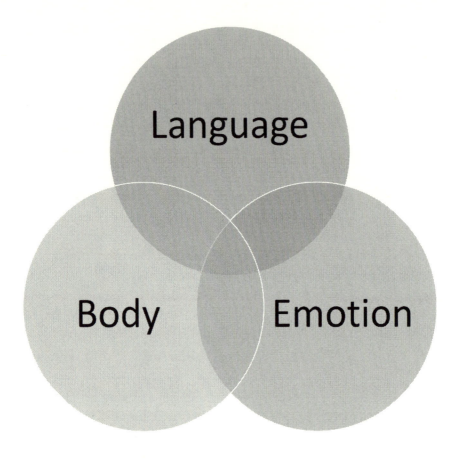

Figure 5.1: Three domains of noticing: language, body, and emotion.

Big Boys & Girls Don't Feel

Emotions can be a rich source of noticing and, for example, actors are often very good at both feeling and conveying particular emotions, but emotions can be difficult for many professionals to notice and to express. In our workshops, when we are working with people on emotional noticing, we might ask, "What emotions are you feeling?" and the person might reply, "I was thinking, X, Y, and Z." We'll ask again, "No, no. we're really interested in what you're feeling. What was the emotion you felt?" and the person will again respond with what they were thinking, not what they were feeling. This sequence of asking for a feeling and getting a thought in response can go on for some time, so we'll give examples of emotional words: "Maybe you were hurt, or angry, or frustrated, or indifferent. What was it?" and eventually the person will use emotional words to describe their feelings, but experiences like these suggest that emotional noticing can be difficult for many professionals. Earlier we called forth the distinction between technical rationality and reflection-in-action, and we believe the widespread belief that good professional practice is the application of theory to particular situations—what we have called *technical rationality*—is responsible for these reactions. After all, if being good as a professional is to be rational and apply well known and regarded theory, to do otherwise, in a sense, breaks ranks with one's profession. One of our purposes in this section is to break this misconception and give permission to professionals for a full range of human knowing and noticing.

Our Bodies as Rich Sources of Knowing

Our bodies are also a rich—and often ignored—domain for noticing. Our five common senses—touch, sight, hearing, smell, and taste—constantly give us inputs that can be noticed, and other detectors in our nervous system give us signals about our sense of balance, where our limbs are located in space, and various signals of internal well-being and distress. For example, athletes, dancers, and martial artists often have heightened kinesthetic noticing capabilities; musicians often have heightened auditory awareness; visual artists often have heightened visual noticing and awareness. The key thing here is that all these noticing skills in body can be cultivated. Listening to your body can offer insight to professionals whose practice is not obviously aligned with bodily coordination and sensation. By directing attention to our bodies and our senses, we can notice those bodily inputs in ways that may be helpful to the cultivation of different kinds of insight.

Whom (and What) Do We Notice?

Another dimension of the noticing game is the "who" dimension. Whom do we notice? Restricting ourselves for a moment to the domain of linguistic sentient beings, whose thoughts, feelings, and body feelings do we notice? Do we notice our own or those of others? In social settings, our ability to notice ourselves and others is helpful to navigating the setting well.

Beyond other linguistic beings, what else do we notice in our environment? Do we notice animals and other living beings, do we notice things that are aesthetically pleasing? Do we notice natural objects or artificial objects preferentially? Are some objects in our environment more salient than others, possibly because of our professional training?

The point here isn't to suggest some one-size-fits-all approach to noticing and awareness. We simply are calling attention to our attention and to become more aware of who, what, where, and under what circumstances we tend to notice (or not).

Two Models to Help Separate Noticing from Judgment and Action

One benefit of calling attention to noticing is that it helps us separate what we notice from any judgments or conclusions we draw from what we notice. Our brains are interconnected clusters of neurons and when one thing happens in one part of the brain it often easily triggers other clusters of neurons, and very quickly we've "jumped to a conclusion." Organizational guru Chris Argyris (1990) called these rapid chains from human sensory input to action the *ladder of inference* with the following rungs:

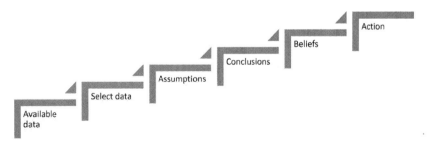

Figure 5.2: Argyris's ladder of inference demonstrates how we jump to conclusions.

Simply stated, our sensors take in available data, and our beliefs and biases then select portions of the data that are relevant or salient. Then, we fill in gaps in the data stream by making assumptions based on past experience or belief, and this leads us to draw a conclusion. The conclusion can confirm or change our beliefs, and then our beliefs and our conclusions can also lead us to take action. The ladder of influence highlights human capacity to be, as one wag once said, "living, breathing assessment machines." And therein lies the problem. The processes just discussed can happen in an instant; we can be calm and peaceful one minute and upset and moving toward a breakdown in another.

By calling out noticing and treating it as a skill that can be developed, we build stronger clusters of neurons early in the chain, and these "noticing muscles" allow us to become more aware more quickly of "the what" we're noticing, and this in turn allows us to take a meta-view and examine the ladder of inference itself, possibly in real time. In this way, viewing our thought process from a distance allows us to modulate or put a brake on the conclusions and actions that would otherwise automatically follow. This is crucial. As we become more skilled noticers, we can more quickly and more effectively catch ourselves in the act of jumping to conclusions that don't serve us, thereby preventing or lessening interpersonal breakdowns and undesirable behaviors.

The ladder of inference is a useful model that helps us understand the microstructure of our thinking. Another metaphor that helps us highlight the importance of noticing is from Heifitz and Linsky (2002). They suggest that increased awareness and noticing allows us to *go to the balcony* and look *down on the dance floor*. In this take, improved skill as a noticer gives us the ability to distance ourselves from the dance of life we're in, thereby permitting cleaner, clearer thought and more appropriate action.

Whether you like the model of the ladder or the metaphor of the balcony and the dance floor, thinking of amplified noticing in these ways helps us continue to build our skill as noticers in ways that serve us as actors and practitioners in the world.

> **Reflexercise 5.2:** In the previous reflexercise, you jotted down some notes about what you noticed. Having just read the section on noticing, what do you notice about your noticing, in general? Do you prefer noticing your thoughts, your feelings, or your body sensations? Do you tend to notice yourself or others most often? What else do you notice about your noticing?

Listening

We started the chapter by calling out curious listening as a keystone habit central to high-performance educational change, and in the previous section we spent some time considering the noticing and awareness part of tri-fold curious listening puzzle. This may seem odd, as it seems that listening is listening and noticing is noticing, but a key point we are making here is that the kind of listening required for deep educational (and personal) change is composed of a number of delineable skills. In this section, we address the "L" listening part of NLQ by making a distinction between two different kinds of listening, *sharing listening* and *curious listening* or what are sometimes called level-1 and level-2 listening respectively in the coaching literature (Whitworth et al., 2009). As our view is that there is no inherent hierarchy here—listening to share and listening to understand are valid and simply different types of listening. We call them type 1 and type 2.

Listening to Share (Type-1)

The first kind of listening we explore is *listening to share* or type-1 listening. In type-1 listening, we hear the words of the other person, but our internal voice is telling us what the words mean to us. With type-1 listening, we often wish to share our thoughts and will interrupt the speaker and do so. Our attention in type-1 listening is on "me," my thoughts, my feelings, and assessments. The main question with type-1 listening is "What does this mean to me?" This is neither good nor bad, and many life situations are ripe for listening to share. For example, cocktail parties are driven by ping-pong-like volleys of talking and listening to share, and a good time is had by all.

There are, however, situations where listening to share or type-I listening doesn't fit well. For example, the following is an excerpt of a dialog between an engineering student and her faculty advisor [Adapted from *Co-Active Coaching, 2nd ed.* (Whitworth et al, 2009), copyright 2007 by Laura Whitworth, Karen Kimsey-House, Henry Kimsey-House, and Phillip Sandahl with permission of Hodder and Stoughton Limited through PLSclear]:

> **Student:** *The new semester is a disaster. I've got five technical classes, profs who keep piling on homework, and I'm not sure that engineering is even a good fit for me. I really miss drawing and painting like I did in high school. And I've got a big mechanical design assignment due next week.*
>
> **FACULTY ADVISOR***: I went through a similar phase when I was your age. The key is to make sure you've got a clear vision of your engineering career in sight.*
>
> **Student:** *That's sort of the problem, though. I thought the promise of a job and high pay was enough, but if engineering work is like engineering school, I'm not sure I want any part of it.*
>
> **FACULTY ADVISOR***: That'll be OK. Your worries are temporary. Don't let them divert you from the real issues— getting good grades and graduating.*
>
> **Student:** *This feels like more than a little diversion.*
>
> **FACULTY ADVISOR***: I'm sure you can tough it out. I had my share of tough semesters too, and I'm glad I stuck with it. In the meantime, let's get back to your design assignment.*
>
> **Student***: Ok. If you say so . . .*

Clearly, the faculty advisor is listening to the student from his own experience and life, but in so doing, he's missing valid concerns that go beyond the work at hand.

Listening to Understand (Type-2)

Next, we explore what we call *listening to understand*, or simply type-2 listening. In type-2 listening, our attention is focused on the other person. The listener asks questions to understand the speaker and doesn't give advice or relay their own thoughts and feelings. In type-2 listening, the speaker feels felt, and the listener is intent on understanding the speaker's true meaning. The body language in a type-2 conversation frequently expresses mutual connection, and sometimes awareness of the outside world is reduced. The main question with type-2 listening is "What does the speaker mean by this?"

Again, type-2 listening is neither better nor worse than type-1 listening, but there are times when listening to understand (type-2) is superior to listening to share (type-1). Our previous advising example is one of those cases, and let's see how the conversation evolves between student and advisor from the same starting point when the advisor listens to understand [Adapted from *Co-Active Coaching, 2nd ed.* (Whitworth et al, 2009), copyright 2007 by Laura Whitworth, Karen Kimsey-House, Henry Kimsey-House, and Phillip Sandahl, with permission of Hodder and Stoughton Limited through PLSclear]:

> **Student:** *The new semester is a disaster. I've got five technical classes, profs who keep piling on homework, and I'm not sure that engineering is even a good fit for me. I really miss drawing and painting like I did in high school. And I've got a big mechanical design assignment due next week.*
>
> **FACULTY ADVISOR:** *In what ways is art important to you? This is a critical period in your engineering education.*

Student: Art helps me express myself and it helps me keep a sense of balance. Right now, I feel like a bit of a robot.

FACULTY ADVISOR: How can you do art and finish the engineering education you've started?

Student: I suppose I could hire a body double.

FACULTY ADVISOR: I can see this is a real dilemma. You're facing a conflict of values in more than one important area of your life. What can I do to help you work through these issues?

Student: I feel so frustrated and trapped. I wonder if you could help me brainstorm some short-term and long-term options?

FACULTY ADVISOR: I'm happy to help. Let's start by looking at your current schedule and workload.

Notice the almost total lack of advice giving here by the advisor. The offering of advice can be received by the other as a form of one-upmanship and it may thereby be resisted on ego grounds. Notice, also, the prevalence of question asking. We will explore question asking in a moment, but listening to understand and open-ended questions are deeply linked. Finally, notice the advisor's devotion to noticing what the student is saying and feeling. The dialogue is a good example, how noticing, listening (to understand), and questioning (NLQ) work together in the business of curious listening.

Before continuing, we recommend that you read the following exercise, and experience both listening to share and curious listening side by side.

Reflexercise 5.3: Pair up with a friend or colleague. Decide which of you will be the *listener*, and which will be the *storyteller*. If there are 3 of you, the third person is the *observer*.

The storyteller should think of a recent experience in which there was some kind of challenge or small conflict that may still be unresolved (this doesn't need to be earth-shattering drama. Everyday interpersonal or professional challenge will suit the exercise best).

The storyteller will tell the story twice. In the first listening, the listener will listen to share (type-I) and in the second listening, the listener will listen to understand (type-2). Note that because of the different styles of listening the storyteller won't tell the stories in exactly the same way. This is quite normal.

First listening instructions: Storyteller tells the story to listener and the listener listens to share (type 1). Listener interrupts storyteller to share his/her personal reactions, advice, similar experiences or the like in response to what the listener is saying. In some cultures, interruptions may be considered rude, but here it is important to exaggerate listening to share to gain insight. Listeners should share their thoughts and feelings readily.

Second listening instructions: Storyteller tells the same story to listener and listener to understand (type 2). This time listener may not offer advice or relay similar experiences. Listener may ask questions, especially about particular terms used by the storyteller (e.g., What do you mean by "a good fit for me").

Debrief: After the exercise, the storyteller should reflect and share how it felt to be listened with type-1 and type-2 listening. The listener should notice how it felt to listen with type-1 and type-2 listening. If there is an observer, he or she should share what he or she noticed about the difference between type-I and type-II listening. All participants may then share other things that they noticed about the two conversations.

Switch roles and repeat. If you have time, have the storyteller become the listener and the listener become the storyteller, and repeat.

A Key Co-Contrary of Listening

As we reflect on the foregoing examples of listening, it is interesting to notice that together type-1 and type-2 listening manage a key co-contrary: *regard for self* && *regard for other*. Listening to share is primarily about regard for your internal voice and the ways in which the stream of words coming from the other person is relevant to you. Listening to understand is primarily about taking the stream of words coming from the other and working to understand the other person's meaning. Managing *regard for self* && *regard for other* is a fundamental life co-contrary and it should come as no surprise that it plays such a central role in the two types of listening called out here.

Questioning

We have already seen how asking good questions is an integral part of curious listening. Here we consider questions in somewhat more detail. We start by discussing different types of questions, continue by sharing a simple trick for asking powerful questions, and conclude by revisiting the listening exercise of the last section with the addition of the asking of powerful questions. Because of their importance, we will often use the abbreviation Q for the term "question."

Types of Qs

Asking questions is such a commonplace that we rarely consider the types of questions of we ask. Here we consider three types:

1. Information-gathering Qs
2. Leading or assumptive Qs
3. Open-ended Qs

Consider each of these in a bit more detail.

Information gathering questions are quite common, and the goal of the asker is to get some information. "How long have you been in engineering?" "Where were you born?" "Do I have your permission to ask a question?" are all examples of information-gathering questions. In our framework, both curious and sharing listening may lead to important information gathering Qs.

Leading or assumptive Qs are asked from the perspective of—and are usually in the interest of—the asker. For example, an appliance salesman might ask, "Would you like the refrigerator in white or stainless steel?" A lawyer might ask, "When did you stop beating your wife?" The faculty advisor in the first dialog of the last section might ask, "Don't you think it would be better to get back to working on your senior design project, rather than worrying about arty and abstract questions?" Leading or assumptive questions can be of great value in certain contexts, but the primary interests being considered are often those of the asker, not those of the person asked.

While information-gathering and assumptive questions have a fairly restricted domain of answers, open-ended Qs are more expansive in scope, and they often require fairly deep reflection on the part of the answerer, especially if the domain of inquiry is new or unfamiliar.

For example, two of the three Qs asked by the faculty advisor in the second dialog in the previous section were open ended: "In what ways is art important to you?" and "How can you do art and finish the engineering education you've started?"

If one's goal is to facilitate insight in another person, open-ended Qs are on the royal road. This is so important that open-ended Qs designed to help another gain insight are given the special name *powerful questions.* In the next section, we discuss a coaching secret that helps us create powerful Qs in a jiffy.

Q-Secret of Coaching Professionals: Begin with the Word "What"

The easiest way to create a powerful question is to begin with the word "what." Here are some examples of powerful Qs [adapted with permission from Marilee Adams's book, *Change Your Questions, Change Your Life*

(Adams, 2015, Location 1418), Berrett-Koehler, Inc. San Francisco, All Rights Reserved, *www.bkconnection.com*]:

1. *What do you want?*
2. *What are your choices?*
3. *What assumptions are you making?*
4. *What are you responsible for?*
5. *In what other ways can you think about this?*
6. *What is the other person thinking, feeling, and wanting?*
7. *What are you missing or avoiding?*
8. *What can you learn? . . . from this person or situation? . . . from this mistake or failure? . . . from this success?*
9. *What action steps make the most sense?*
10. *What questions should I ask (myself or others)?*
11. *What can turn this into a win-win?*
12. *What's possible?*

These are great generic powerful questions. As you read through the list, go slowly and mark your two or three favorite questions. There is nothing special about the wording of these Qs. For example, we will often ask a variant of Q 11 as "What is success for you in this situation?" Q 10 is a favorite when we don't know what to ask: "I'm a bit stuck and don't know what to ask; what Q should I ask you right now?"

Perhaps the question we ask the most often is simply, "What else?" Prompting someone to continue their train of thought lets them know you're keeping up with them and asks them to keep going.

Why "What" and Not "Why" and Sometimes "How"

Sometimes, in audiences familiar with quality methods in manufacturing, we'll get questions about why we prefer the word "what" to the word "why?" In quality methods, one way to get to root causes is to ask "Why?" five times (the "five why's") in the hopes of getting to an underlying root cause, and this is generally a good approach in impersonal settings.

Unfortunately, in one-on-one personal settings, asking someone "Why did you do X?" is likely to get a response that defends the decision, so "What" questions that surround the incident can draw out both what the person did, what their motives were, and what they learned from doing it: "What was involved in doing X?" "What were your motives in doing X?" "What did you learn from doing X?"

"How" Qs are often used to understand processes or the ways in which something was accomplished or can be accomplished. "How" Qs are fine for finding about the past, but for uncovering future approaches, we notice that sometimes a simple "How" gets a single response (the "one right way") rather than a reflection on different ways the think can be done. Instead. The locution, "In what ways can you approach X?" can get someone thinking beyond a single approach.

Mixing It Up with Prompts

Sometimes, asking a string of questions can become too repetitive, so mixing it up with an open-ended prompt or two can work quite nicely. Here are some sample Q-to-prompt substitutions:

- "What else?" → "You also were thinking . . ."
- "What action makes the most sense?" → "The action that makes the most sense is . . ."
- "What are you missing or avoiding?" → "You are missing or avoiding . . ."

In conversational settings, "what" Qs are always a good bet. Sometimes prompts work well when you're asking someone to write something, and the first few words prime the pump for their answer. Practically speaking, either open-ended Qs or prompts can be used relatively interchangeably.

Reflexercise 5.4: Repeat reflexercise 5.3 by (1) skipping the type-1 listening portion, (2) having the listener tell the same or a different story, and (3) having the listener ask a series of powerful whatQs as they listen.

Debrief. Storyteller and listener should debrief as before. Also, storyteller and listener should notice which questions elicited the most powerful responses.

Switch roles and repeat. If you have time, have the storyteller become the listener and the listener become the storyteller, and repeat.

Two Q-Habits to Avoid: Q-Stacking & Q-Perfectionism

When asking questions, simply ask one at a time. Sometimes you'll hear radio or TV interviewers stack two or more questions (Q-stacking). Worse, in the media the Qs themselves are often leading or assumptive Qs. This sometimes comes from the interviewer wanting to control the outcome of the interview, but when you're helping someone have insight through curious listening, you're really interested in what the other person has to say. Moreover, asking multiple questions is confusing. Just ask a question and let the person speak. If the second Q in the stack still makes sense after the response, you can still ask it, but more likely some other question will rise to the surface and be more important than your previous Q.

Another problem early in powerful Q training is the idea that one must ask the perfect what Q to elicit a good response from the client. This simply isn't so. Oftentimes you'll ask a question, and your client answers a different Q entirely, the one they really wanted to answer. In these cases, your Q was a prompt, but wasn't all that important from a content perspective. Sometimes Qs do land really well and over time, you'll learn how to hit the mark more often, but the idea that your question is central to the other person having an insight comes from the same place of ego as listening to share. If the other person is resourceful, creative, and whole, your role in the conversation is more of letting the other person blossom and express themselves. Sometimes a Q doesn't land all that well. So, you get a chance to ask another one. Just ask one, and let it go.

> **Reflexercise 5.5:** Journal or make some notes about your takeaways and action-aways from this chapter.
>
> **Some questions.** Skim back over this chapter. What parts resonated with you? What parts did you disagree or have some concern with? What terms were used in ways that seemed familiar or unfamiliar? What experiences in your life connected to what you read? What questions arose in your mind for further reflection and consideration? What inspired you to action?

Dip or Dive

1. **Emotional intelligence and noticing.** *Dip.* Watch the short video on emotional intelligence by Daniel Goleman (Big Think, 2012) and pay attention to role of noticing of self and others in it: *https://www.youtube.com/watch?v=Y7m9eNoB3NU.* Journal about what you notice. *Dive.* Read either of the texts *Vital Lies, Simple Truths* (Goleman, 1985) or *Emotional Intelligence* (Goleman 1995) and repeat.

2. *Change your questions, change your life.* **Dip.** Watch the short video (Berrett Koehler, 2008) *https://www.youtube.com/watch?v=Ahy4WfnA5AM* by Marilee Adams discussing the content of her book *Change Your Questions, Change Your Life* (Adams, 2015). In what ways are learner questions different from judger questions? What else do you notice from the resource? *Dive.* Get a copy of the text and repeat.

Reflectionaire

1. **Emotional words.** From experience or memory, make a list of words that describe emotions. Use a search engine (search for "emotional words") to find expanded lists of emotional words. What do you notice about your list and the expanded list? Of the emotional words that you have uncovered, what are your top 2-5 emotional words in terms of frequency? What are your top 2-5 emotional words in terms of intensity? What else do you notice as you go through this exercise?

2. **Body sensations.** Sometimes our bodies talk to us and give us gut feelings of either warning or attraction. When your body gives you a good or bad gut feeling, where or in what different ways do you experience it? What variation in body feelings do you have for bad or foreboding feelings? What variation in body feelings do you have for good or creative or generative feelings? What else do you notice about body feelings?

3. **Noticing avoidance.** What things do you avoid noticing? What things do you regularly notice poorly or incorrectly? What experiences in your past might account for this twist in your noticing circuitry? What emotional and body feelings come up for you as you do this exercise?

4. **Noticing others noticing.** Go through a week and heighten your awareness of the degree to which others are noticing or aware. Attempt to do this without judging the others. Make some notes or journal about this activity. What things appear to compete for the attention of others? What else do you notice about this activity?

5. **Noticing your noticing.** Repeat reflectionaire 4 above for yourself. Be gentle with yourself and do not judge your noticing. What patterns do you notice in your noticing? Whom do you notice most (and least)? What do you notice most (and least)? When do notice more (or less)? What emotions and body feelings come up for you as you go through this exercise?

6. **Listening to Qs.** Listen to a radio, podcast, or TV interviewer and notice their questioning technique. What kinds of questions do

they ask? To what extent do they stack Qs? To what extent do the interviewees answer the question asked or give an answer that is in some ways non-responsive or not aligned with the question? What else do you notice in your listening?

7. **Listening to listening.** Notice your conversations with others over several days. Without judging others, to what extent do you believe the people listening to you are in type-1 or type-2 listening, what proportion of the time? Without judging yourself, to what extent do you believe you are in type-1 or type-2 listening what proportion of the time? What else do you notice in your conversation noticing?

8. **DIY (do it yourself).** Upon reading this chapter, what concept, experience, innovation, or other object caught your attention that deserves further reflection? What is it about this object of your attention that interests you? What have you experienced in connection with this object? What other questions need to be asked and reflected upon in connection with your DIY reflectionaire? Reflect on those questions. What else do you notice as you reflect upon this object?

Little Bets

1. **NLQ in classes with collaboration or teamwork.** As a faculty member or a student leader of extracurricular activities, run the curious listening-NLQ reflexercises of this chapter as team-building exercises to increase mutual listening and better team coordination. Use the language of curious listening and NLQ to reinforce the use of these methods. At the end of the semester or activity, debrief the extent to which these methods were helpful or not.

2. **Intentional noticing without intentional change.** Consider some aspect of your life that you would like to change. Without judging yourself (be gentle) and without intention of changing, simply direct your attention, your noticing, to the behavior for one week?

Simply journal about what you notice. Without intending to change, to what extent did the behavior change once you directed attention to it? To what extent did it remain the same? To what extent were you able to notice with or without judgment? What else do you notice?

3. **DIY little bet.** What little bet needs to be made as a result of your thinking about the material in the book so far? What is easiest to try or do with existing means? What is the smallest possible experiment that will get you some interesting results? What do you notice from doing this experiment? What else needs to be done?

6

Four Sprints & Spirits for Rapid, Embraced Change

When we wrote WNE it was our shared experience and belief that educational change is quite slow, taking years to get a typical faculty to agree to make substantive changes to required courses in a given department or unit, but this experience and belief conflicted with our observations about how quickly you can get cultural change in a unit working together in a good mood in a collaborative way. The Olin bootstrap, the iFoundry startup, and other similar experiences showed us the tie between mood, culture, and change in ways that opened the doors to a different way of thinking about the speed of embraced change. Then, not long ago, one of us (Goldberg, 2019a) was approached to help facilitate a curriculum change process in an engineering department in South America. The client was asked, "how many years do you have to make the changes?" and the client replied, "two months." Despite doubts about the likely success of the endeavor, a project began entitled Four Sprints for Rapid, Effective Curriculum Change. This pilot project has led us to put forward a template for educational change that integrates the lessons we learned in writing WNE and since.

Speed with Embrace

Of course, speed of change is important, but one of the easiest ways to get rapid change is to have a dictatorial leader who is willing to make change through fear and coercion. The problem with this kind of change is that people will appear to go along with it if the coercion is sufficiently forceful, but this kind of change is rarely *embraced*. The process of this chapter is the opposite of fearful and coercive. In many ways, it is caring and loving, and as the process continues, the conversations increasingly become comfortable with the ideas and particulars of change; they are increasingly embraced. Although our work in WNE and since has largely been in an engineering context, we believe the processes of this chapter apply to relatively rapid, embraced change in major initiatives throughout the academy.

In this chapter, we start by discussing why the typical approach to educational change fails. This leads us to the approach of the field manual, one that use four sprints and four spirits together with the five shifts of chapter 4. Together we call this the *four sprints* & *spirits method* or simply 4SSM. In particular, the chapter discusses each of the sprints, each of the spirits, and the ways in which both interact with each of the shifts.

Educational Change as Mission Impossible

Academic leaders or faculty who try to make substantive change almost immediately feel like agent Ethan Hunt taking on yet another impossible mission, especially the part, "As always, should you or any of your IM Force be caught or killed, the Secretary will disavow any knowledge of your actions," and the difficulty of educational change should not be underestimated. Having said this, we believe an integrated, humble, and principled approach to change can bring about movement in months, not years.

To understand how to make rapid, embraced change, we start by listing eight reasons why traditional educational change fails:

1. It is ego-driven arm-wrestling
2. exacerbated by faculty research factions
3. resulting in a kind of NIMBY problem
4. with deliberations performed by committees without facilitation
5. that ignore the importance of cultural and emotional shifts
6. and merely shuffle and tweak existing course boxes
7. largely ignorant of innovation elsewhere
8. oblivious to major stakeholders, especially students.

The first three of these points are crucial. Change committees are filled with big egos and the arguments for this or that curricular element, after clearing away a lot of bluff and bluster, are no more sophisticated than, "the best curriculum is the one I took as an undergraduate" (item 1). This difficulty at the level of individuals is exacerbated by a group dynamic in which factions form (item 2) around faculty research interests ("Our research specialty is essential for the education of students in this department."). Both the individual and factional ego problems lead to a NIMBY = "not in my backyard" problem (item 3) like the siting of a nuclear power plant, in which people want the power but not the plant. The educational equivalent is, "Innovation is great, just don't change my course." The factional equivalent is, "Innovation is great, just don't mess with my faction's required course share."

These problems are compounded by poorly structured processes (item 4), a lack of understanding of cultural and emotional shifts (item 5), an overemphasis on current architecture and coursework (item 6), a stunning lack of awareness about the best practices at home and elsewhere (item 7), and lack of involvement of other stakeholders, particularly employers and students (item 8).

> **E-Box: Factions, factions, everywhere.** In a midwestern public university in the US, a longstanding engineering department has faculty members in three main research factions: control systems, structural and machine design, and operations research. Over twenty years, more and more of the curriculum is devoted to each of these factions getting a bigger piece of the required course pie, with less and less time allowed for student electives or choice. This occurred at a time when it would have made sense to make sure that students had the option to take a bigger dosage of engineering computation and/or product design and management coursework. Not once in that 20-year period were students asked what they wanted from their degree or the ways they found it to be relevant, old-fashioned, or just out of touch.
>
> **Lesson.** Faculty often argue amongst themselves for their own ego and research interests, but student interests are rarely consulted, let alone included in the larger educational design conversation. Higher education would be better with even modest improvement in this regard.

As we think about ways to overcome some of these difficulties, the arm-wrestling, factional, and NIMBY problems loom largest, and a key element of any effective method is to engage change team members with key theories and practices with which they may not be familiar. This is a central concept in the 4SSM approach, and the four sprints and spirits combined with the five shifts take the change team members on an engaging journey of discovery and conversation. These conversations almost immediately shift attention away from Dr. X's curricular experiences at HiTechU and Prof. Y's glory days at UMagnificent with everyone asking good questions about what is needed in our program and why.

The idea that traditional university committees can be used to bring about significant change is another problem. Committees are often populated using the Noah's Ark principle ("two of each animal") and as a result, finding consensus in weekly or monthly meetings is almost impossible. Between meetings, adherence to commitments is often spotty, and at the end of the semester the committee writes a report to brag about (or put lipstick on the pig of) all that it has "accomplished." To be more effective than this, it is helpful to go against the grain of traditional committee governance and bring together a motivated group of principals to work in structured, stagewise fashion with discrete deliverables, clear and differentiated accountability, a clear finish date, and the insistent and repeated expectation of concrete results.

Moreover, attempts to change without proper facilitation in these conversations is also a key difficulty. Because of the ego attachment to particular courses and solutions, meaningful conversation about these topics requires a degree of conflict that many faculty members will reflexively avoid. A skilled facilitator welcomes the conflict and sees it as essential to the subsequent embrace of change. Given the importance of conflict, it is also important that the facilitator be neutral to any particular change. A key here is the idea of *humble* facilitation (Schein, 2013) in which the facilitator acknowledges different points of view and does not know what the change team should do. Thus, facilitation by a knowledgeable and neutral facilitator will allow change principals to have the conversations that need to be had in ways that allow differences to be aired, acknowledged, and resolved or set aside, as appropriate.

The tendency to shuffle and tweak existing course boxes goes together with ignoring the possibilities of cultural-emotional shifts and the usual ignorance of innovations elsewhere. Typically, the status quo curriculum is the starting point for all conversation, and opportunities to innovate outside the curriculum in the culture and mood of the place are ignored. Moreover, many faculty members will be unfamiliar with innovations, even in their own department, faculty or college, and university. By starting with noticing innovations at home, we raise awareness that substantial change can occur, in fact, has already taken place. Looking at innovations outside the school and around the world inspires changes that might not otherwise have been imagined or possible.

Reflexercise 6.1: Consider one or two examples of curriculum reform or educational change that you have been a part of or otherwise have experienced. Reread the eight failures modes listed above.

Reflection. To what extent were the change efforts successful? To what extent were the change efforts more or less successful than desired? In what ways did the successes avoid or overcome the failure modes listed above, and in what ways did the less-than-successes succumb to the failure modes listed? What else do you notice as you reflect on the examples and the success and failure modes?

Four Sprints & Spirits Method as a Hybrid of Three Action Frameworks

The design of the four sprints and spirits method takes three powerful action frameworks and brings them together to form a powerful hybrid:

1. Agile development
2. Design thinking
3. The five shifts

The idea of a *sprint* is borrowed from agile software development (Atlassian, 2021; Hills & Johnson, 2012). In agile software development, small action steps are taken in a fixed time period—the *sprint*—with key deliverables required at the end of each sprint. The aggregate of repeated sprints is the accumulation of significant action and learning.

4SSM borrows from *design thinking* the notion of a studying your customers or users like an anthropologist, learning how they work and live and what their wants and needs are. This results in the creation of several discrete and concrete customer types, called *personae* (Innovation Training, n.d.). In design thinking, personae are used as a constant reminder of who your customers or users are and what they really value, want, or need.

Finally, the five shifts approach to practical insight of the last two chapters is used as the underlying reflection-conversation operating system of the 4SSM. Each of the shifts contributes to a whole-person approach to designing educational programs for students and in this way, sprint-by-sprint, persona-by-persona, and shift-by-shift, the change initiative is led to good educational designs.

4SSM Team Structure and Other Roles

The first move in 4SSM is to replace the usual haphazard committee-based "change" process with a more structured process with a change team and several well-defined roles:

- Change team (CT)
- 4SSM facilitator
- Scribe/data coordinator

Each of these is discussed below.

The Change Team (CT)

The change team is a group of individuals, usually faculty and other stakeholders, possibly including students, alums, employers, and others, assembled to design the educational change. In selecting CT members, it is important to gauge both their motivation for team membership as well as their predisposition toward change.

Team Selection

This is not to say that all members need to be all-in for change prior to joining the CT. Different individuals have different predispositions towards change in general. While it's not necessary or even desirable to recruit full-fledged revolutionaries, it is important to gauge whether people are implacably invested in the status quo.

E-Box: How not to select a change team. In one change engagement at a major public university in the US, the dean provided a significant financial incentive for faculty members to join the change team, and as you might imagine, a number of members signed up for the money not for the change. In that case, it took over a year to sort through those whose motives were aligned with the project and those who simply lined up at the latest feed trough.

Lessons learned. When possible, gauge authentic interest by getting individuals to make some small sacrifice to be involved. Have them attend meetings, commit to small work assignments, or otherwise show genuine interest in the project, prior to their being selected to join the team.

One of us has written elsewhere about the three speeds of change (Goldberg, 2015) as being *SLOW, GO, & NO.* Potential CT members who are cautious and have significant concerns with the risks of change (the SLOWs) are very welcome because they can help the group manage the co-contrary between change and stability. Potential CT members who have been chomping at the bit for change and have concerns with the risks of not changing (the GOs) are also very welcome, and they are often the pistons driving innovations. Those who are adamantly opposed to change (the NOs) can be problematic as team members, especially when they defect from group norms of fair procedure.

This can especially be a problem when the CT is chosen as a whole department or unit. In such cases, some NOs are inevitable, and sound process is crucial. Given some time, the logic and the conversations built into the 4SSM method will soften even the most steadfast NOs somewhat. A key to continuing in the face of stiff resistance is to (1) get the group to accept the process as the way forward and (2) trust the process. If the process is accepted by the group, those who resist or sabotage the process should be removed on fair process grounds; when they are removed, they should be told that they will be welcomed back, if they accept the process.

Change Team vs. Departmental Leadership

One important issue is the relationship between change team leadership and the titular leadership of the department or unit. Oftentimes in a change initiative, there is an individual or group of individuals that are the sparkplugs for the initiative. These sparkplugs may or may not be titular leaders of the unit. When CT leaders do not include members of the unit's formal leadership team, it is important to have or earn the tacit or explicit support of the unit's leaders, lest the whole effort fail on political grounds. As the process continues, other leaders emerge in the various sub-divided activities of the change team.

On the other hand, when unit leadership is pushing for change, it is important to lead with a light hand. Too heavy a push in too directed a fashion will close off discussion and debate. A key element of the process is respect for the co-contrary of *tradition && innovation*, so change-oriented leaders can push the process, not any particular agenda, and allow good stuff to emerge from the reflections and conversations built into the process.

Team Charter

One of the first things the team should do when coming together is to write and commit to a team charter that specifies how frequently the team will meet, what responsibilities members have to each other, how long the charter lasts, what commitments the team is making to the process, what rules and norms including attendance and participation, and how the team will lead itself. Taking time at the beginning to discuss and agree to a charter will prevent problems late in the process.

The Facilitator

Hiring an independent 4SSM facilitator improves adherence to process and more importantly improves the likelihood of substantive outcomes. The 4SSM facilitator is a person with experience in coaching, organizational development, or educational change who can facilitate 4SSM engagements using the materials, exercises, and facilitation methods of the process.

The facilitator is not a leader or part of the CT, but rather he or she is someone who visits virtually or in person with the CT several times before, during, and after each of the four sprints. Facilitators are entitled to their own views about what might or might not be changed in a given engagement, but it is important to check these opinions at the door. Doing so requires a kind of humility "to not know" what the unit should do or change. In this sense, the facilitator is not a smarty pants know-it-all vis-à-vis change; rather he or she is a guide or Sherpa to the possibility of change, and the more neutral and humble the facilitator is, the more successful the change effort will be.

The Scribe/Data Coordinator

The 4SSM generates a lot of conversation about needs, goals, alternate experiences, and innovations. It is important to capture the key results of these conversations as they are generated and this requires a person whose job it is to record, organize, and preserve the record of the method. As with the facilitator, it is important for the scribe to "not know" what the right answer is and to record the sentiments of the design team members faithfully and accurately. When the process is run in person, taking pictures of posters and sticky notes is a good way to capture session outputs. When sessions are mainly conversational or online, the scribe takes notes or minutes, trying to capture the key phrases and ideas quickly and accurately.

Each of the four sprints begins with virtual or real visits from the facilitator and each spirit sub-team is assigned tasks appropriate to the sprint. In addition, just-in-time training exercises are used at the beginning of each sprint to provide the needed skills for the sprint, to help improve communications, and ready the DT for the change negotiations they face at the end. Each sprint ends with each of the four spirits presenting their results from the previous sprint. During each sprint, the facilitator is available for office hours or scheduled coaching calls to help the CT overcome challenges or obstacles that arise during the sprint.

With these team and structural elements described, we turn to the guts of the 4SSM process, the four sprints themselves.

The Four Sprints

Rapid, embraced educational change is facilitated by holding four sprints in sequence followed by a process of ongoing implementation and innovation:

1. Four spirits of change sprint
2. Bright spots and great possibilities sprint
3. Educational model canvases sprint
4. Negotiating tradition and change (from interests) sprint

The four sprints are depicted in sequence in figure 6.1.

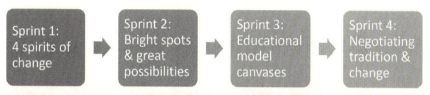

Figure 6.1:   The four sprints lead step by step to a plan for substantive change.

Let's look at each of these in more detail.

Sprint 1: Four Spirits of Change

Sprint 1 introduces the 4SSM process to the change team and holds a number of preparatory team reflections about what kind of change might be beneficial. It holds meetings with key stakeholders to get important data to guide the process. It also subdivides the change team into four sub-teams, each with a particular focus. Let's consider each of these in turn.

Introduction to 4SSM and Preparatory Reflections

Sprint 1 starts with an introduction to the overall 4SSM change process, some preparatory group exercises on students today and the possibility of student unleashing as a source of educational motivation.

The 4SSM introduction contrasts the usual committee-based change process and 4SSM, emphasizing the eight failure modes of committee-based processes discussed above. It also foreshadows the four sprints and gives high-level reasons for their existence.

I-Like-I-Wish Sessions

Additionally, as part of the first sprint, key stakeholders—faculty, students, alumni, and employers—are interviewed in group sessions called *I-like-I-wish sessions*. Sticky notes (or the virtual equivalent) are handed out, and members of each group are prompted to fill out the notes with "I like . . ." and "I wish . . ." about those things they like about the current program and those things they wish were different.

E-Box: "Students today are not . . ." & stories of engaged students. One of the favorite introductory reflections in our workshops is the "Students today are not . . ." exercise. We send faculty members off to reflect about the ways students today aren't as good, as smart, as serious, as whatever as students of yesterday, and the tables get a buzzing with this one. After the debrief, we flip the switch and tell stories of very engaged students (usually *outside the curriculum*) and all the amazing things they do. The contrast is a startling one, and we ask faculty to consider the mood and cultures of environments where students shine and those where they don't.

Lesson learned. These reflections help raise awareness that students are often disengaged for cause, the result of impoverished educational environments.

Quote without comment. When we flip the 2nd prompt from students to faculty, students tell us that "Faculty today are not . . ." *listeners, motivating, flexible, open minded, committed, passionate, prepared, aware of new subjects and technology, or paying attention.*

"I-like-I-wish" is a simple exercise, but in very short order it sizes up key druthers of different stakeholder groups. Once the raw data is received, it should be summarized into salient points for both preservation of the valued elements of the status quo as well key possibilities for improvement and change. In collecting this data, it is important to use the raw terms used by session participants. Doing so keeps the data authentic. Professional facilitation of these sessions helps to get the most out of them quickly and well.

> **E-Box: Students I-like-I-wish.** It's hard to generalize, but when we do "I-like-I-wish" with students they **like** flexibility, rigor and challenge, energetic faculty who prepare for class, project-based learning especially with companies, and a variety of experiences that lead to work. Students **wish** for better teaching of soft skills, more flexibility and choice in disciplinary requirements, more realistic pre-requisites and work/grading requirements, and more up-to-date courses that keep up with timely subjects.
>
> **Lesson learned.** Students have a well calibrated sense of the good stuff (and bad stuff) and how to make it better if we would only talk and listen to them.

The Creation and Activation of the Four Spirits of Change Teams

A key feature of sprint 1 is the creation of *four spirits of change* sub-teams, and it sets the tone for a much different kind of conversation from the ego-driven arm wrestling of traditional educational change. In this sprint, we divide the full change team into four specialty teams, the *four spirits teams*, as shown in figure 6.2.

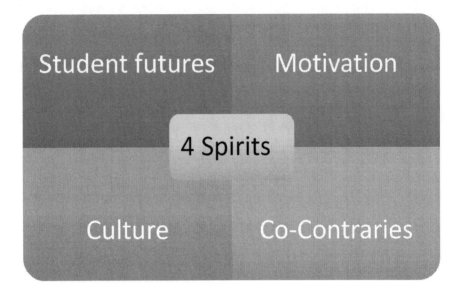

Figure 6.2: The change team is divided into four spirits teams, each with a different specialty.

By separating the change team into four discrete units and asking each of them to become "expert" in a particular facet of change, we redirect faculty energy away from the usual ego affirmation ("my education was better than yours") and toward substantive matters that will help the combined team design something appropriate and good. Notice that the subjects are not particular changes or a particular course of action, but rather meta-level subjects that help faculty understand how organizations learn best.

The Rationale for the Four Spirits

The four spirits (figure 6.2) are not written in stone, and different change projects might choose a different number of facets or different substantive facets appropriate to a particular change task. For curriculum change, the four spirits shown—student futures, motivation, culture, and co-contraries—help the team examine the existing program from several complementary, useful lenses that enable great conversations about what's working now and what needs to be better. In doing this work, faculty say things like, "I always wanted to have this kind of conversation with my colleagues, and now we are." Students say things like, "It feels important for students to be a part of these conversations because we have a different perspective on what our education needs to be than faculty members even a little older than we are."

The first spirit, the student futures spirit, borrows methodologically from the discipline of *design thinking,* specifically using the notion of *personae* (Innovation Training, n.d.) to help conceive of the kinds of students who will benefit from going through the educational program being redesigned. These personae (see an example in figure 6.3) are concrete, have names, have personal life stories, and are designed to reflect both the kinds of students the program has now and the kinds it would like to have if the change initiative is successful. Imagining the kinds of students in concrete terms helps the team design for more effective change.

In the motivation spirit, the team studies short videos and writings from authors such as Dan Pink and Carol Dweck as discussed in WNE (Chapter 6) to understand modern motivation and mindset theory applied to the unit undergoing change. By analyzing the extent to which a department utilizes *intrinsic && extrinsic motivation* in harmony, one can find gaps and ways to balance an overly determined curriculum with one that offers space for student autonomy and expression.

The culture spirit team is asked to become familiar with Schein's theories of organizational culture (Schein, 2009) and apply them to understanding the existing culture of the unit. Chapter 2 of this book and Chapter 8 of WNE briefly discuss these ideas. Examining the educational unit and its culture dispassionately is particularly helpful to understanding why it accepts or rejects certain ideas or approaches.

The importance of co-contraries was examined in some detail in chapter 3 of this book. The spirit team for co-contraries is asked to list key co-contraries affecting education in the unit and determine the degree to which excellence is obtained on both contraries. This exercise is particularly helpful to uncovering gaps in coverage.

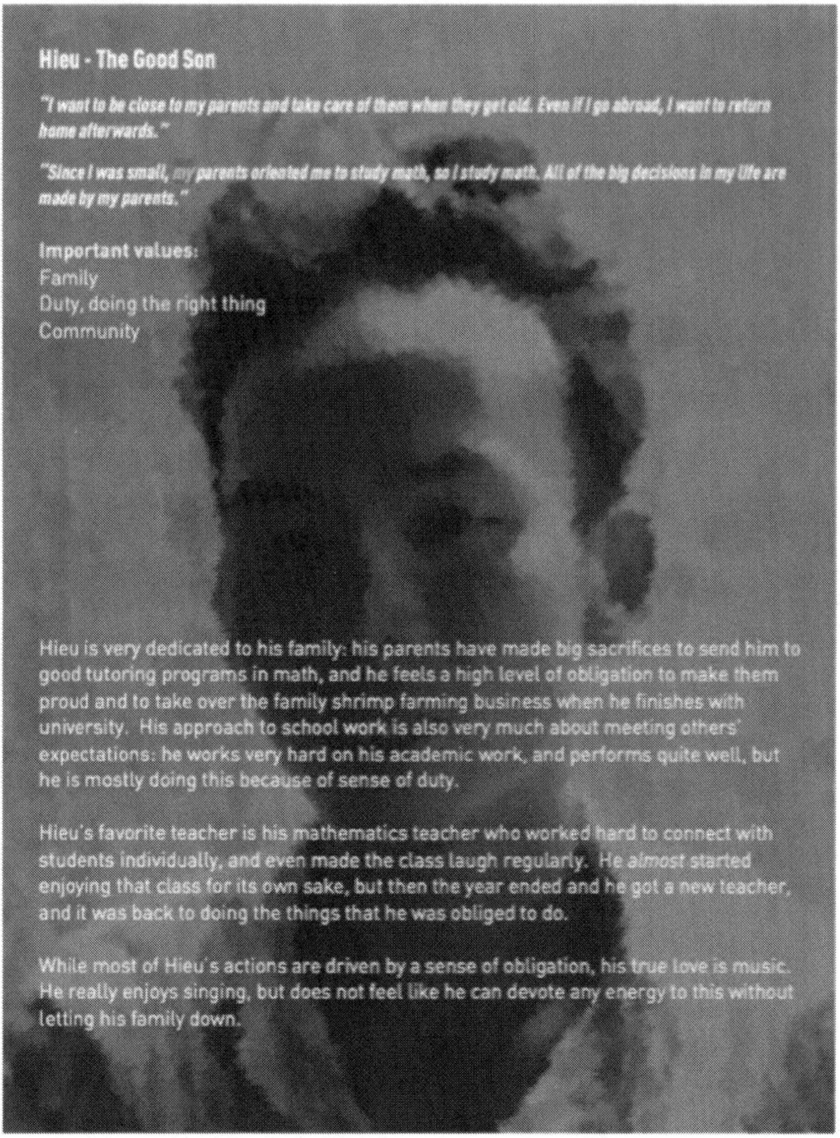

Hieu - The Good Son

"I want to be close to my parents and take care of them when they get old. Even if I go abroad, I want to return home afterwards."

"Since I was small, my parents oriented me to study math, so I study math. All of the big decisions in my life are made by my parents."

Important values:
Family
Duty, doing the right thing
Community

Hieu is very dedicated to his family: his parents have made big sacrifices to send him to good tutoring programs in math, and he feels a high level of obligation to make them proud and to take over the family shrimp farming business when he finishes with university. His approach to school work is also very much about meeting others' expectations: he works very hard on his academic work, and performs quite well, but he is mostly doing this because of sense of duty.

Hieu's favorite teacher is his mathematics teacher who worked hard to connect with students individually, and even made the class laugh regularly. He *almost* started enjoying that class for its own sake, but then the year ended and he got a new teacher, and it was back to doing the things that he was obliged to do.

While most of Hieu's actions are driven by a sense of obligation, his true love is music. He really enjoys singing, but does not feel like he can devote any energy to this without letting his family down.

Figure 6.3: Example of a student persona. Reproduced with permission of Mark Somerville.

Although each of the four spirits teams approach the problem of change through a different lens, it turns out that these perspectives are complementary and lead to good conversations about what might be changed and how those changes might affect different aspects of the unit following the changes. Having a multi-perspective, sophisticated conversation about what might be changed and how it might affect the unit is exactly the kind of conversation that is necessary in an embraced change process.

> **Reflexercise 6.2:** Think about an educational institution with which you are familiar. One by one, quickly make some notes about each of the 4 spirits from your own perspective and existing knowledge. For what future was the institution preparing students? In what ways did this institution motivate students? What salient aspects of the educational culture come to mind? What co-contraries were managed in a balanced way or not?
>
> **Some questions.** As you reflect on these quick observations, what parts of the status quo are important to preserve? What parts are important to change? What else do you notice as you reflect on these matters?

Sprint 2: Bright Spots and Great Possibilities

Sprint 2 is all about being innovative and inventive; it starts by asking each of the four spirits teams to identify *bright spots* (Heath & Heath, 2010)—existing best practices—within the department, in the larger school, and outside the school around the globe. A key to a successful second sprint is to seed the teams with good suggestions of different schools and innovative practices and to require them to explore those or others on their own.

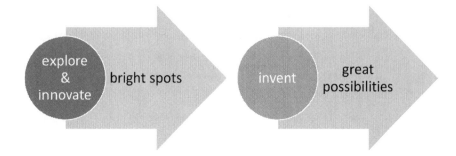

Figure 6.4: Sprint 2 looks for bright spots and great possibilities or candidate elements of a new program.

After taking a good look around, the teams are required to design *great possibilities*—candidate innovations or inventions for possible change in the unit. These can be created using modified bright spots or by inventing new possibilities through brainstorming and creative practice. It is often the case that particular bright spots are copied or cross-pollinated with other good ideas to create a great possibility. Other times, it is the invention of something inspired by some other source or mechanism entirely.

One key to raising energy in this phase is to encourage teams to give their great possibilities labels or names that are sticky in the sense of the Heath Brothers text, *Made to Stick* (Heath & Heath, 2007). Language is sticky if it is a SUCCES, that is (1) simple, (2) unexpected, (3) concrete, (4) credible, (5) emotional, and (6) a story. Sticky language doesn't sound like a big deal, but using imagination to create interesting names rarely fails to lift motivation and mood in a design team.

Reflexercise 6.3: What activities in your program are worthy innovations? What activities elsewhere in your faculty (college) are worthy innovations? What activities elsewhere in your school (university) are worthy innovations? What programs similar to yours, are doing the most innovative things? What are their key innovations?

Some questions. In what ways can the innovations you listed be used or not in your program? What innovations are interesting, but not practical? What difficult or impossible thing could you do to make them practical? What else do you notice?

Sprint 3: Educational Model Canvas

The third sprint takes the innovative pieces of the second sprint together with existing practices and organizes them into a coherent whole. In particular, the *educational model canvas sprint* requires that each of the four spirits teams create at least one educational model canvas (see Figure 6.4). The idea of using a canvas, essentially a large poster, to capture the key elements of an educational proposal is drawn from the well-known book, *Business Model Generation* (Osterwalder & Pigneur, 2010), which used the idea of a business model canvas to integrate conceptual fragments of a new business into a full-fledged business model. Here we do likewise for an educational proposal, although we modify the elements of the canvas to be more appropriate to educational program innovation and thought.

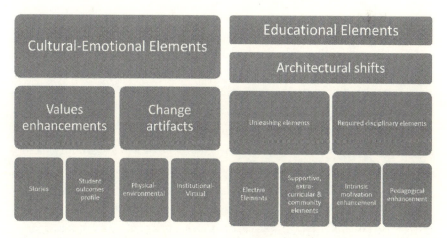

Figure 6.5: Sprint 3 creates several educational model canvases to integrate the pieces of a new curriculum or initiative.

In the 4SSM notion of an educational model canvas, we make the distinction between cultural-emotional elements and educational elements, discussing each category further by parts.

Cultural-Emotional Elements

The canvas starts with the cultural-emotional elements of the program, and this might seem weird—after all, by sprint 3 everyone is chomping at the bit to fight about who gets to teach which required courses—but two key lessons of WNE were that (1) the critical variables of change were emotional and cultural and (2) a lot of innovation can take place without changing a single class. A particular virtue of putting cultural-emotional elements first is that making substantial change in such things as espoused values, sticky language, departmental signage, websites and literature, physical and virtual layout and the like receives much less resistance than even small changes in courses and curriculum content.

Values Enhancements

In the canvas template of figure 6.5 we break the cultural-emotional elements into two big categories, *values enhancements* and *change artifacts*.

Recall in Schein's (2009) 3-level model, he distinguished between artifacts, espoused values, and underlying assumptions. Here we start with values, and specifically we are interested in changes in values or changes in how the department articulates and communicates them. If you collect departmental descriptions, websites, handout literature, and advertising, you'll have a nice package of specifically how the unit describes itself currently. During sprints 1 and 2 there are usually lots of good discussions about the unit, its faculty, its students, its identity, and its aspirations. So, given this deep and meaningful conversation, the values part of the canvas should capture the unit's current and aspirational espoused values going forward. Additionally, the conversation to this point may have come up with good sticky language to articulate the department's values, old and new, and the values box should capture specific memorable ways of articulating the values package to students and stakeholders.

Stories

The sprint discussions are also typically story laden, and the stories box of the canvas is designed to capture key stories that can be told to convey the deeper sense of the unit's mission. Telling stories is a fundamental way in which human beings deal with complexity, and in the stories we tell we integrate rational, emotional, and even body feelings in ways that other forms of communicating do not. Given the rise of YouTube and other video-sharing sites, you can think of this section of the canvas as little video vignettes that will be used to attract, motivate, and energize prospective and current unit members.

Student Outcomes Profile

An important part of the canvas is the *student outcomes profile*. This is a list of desired knowledge, behaviors, and practices that a student in the department should be expected to learn and be able to practice as a result of their education. The unit probably already has these kinds of lists from past curriculum exercises, and these can be updated by reflecting on the "I-like-I-wish" data gathered in sprint 1. Some care should be taken to make these lists at a fairly high of abstraction. This is not the place for a particular research faction to lobby for the importance of three classes in "left-handed quantum physics;" however, it is an appropriate place to call out the importance of "rigorous and creative use of key disciplinary knowledge and practices" or the like.

Change Artifacts

Another box in the cultural-emotional side of the canvas is the *change artifacts* box split into two types: physical-environmental and institutional-virtual. Part of making Olin College special was the design of the physical space that paid special attention to laboratories, collaboration rooms, small studio classrooms, and minimized large lecture halls. Physical changes to the physical space and campus environment can send important messages of seriousness around a change initiative.

Additionally, there are *institutional-virtual* change artifacts that can be helpful to signaling serious change. Creating new signage about change, renaming courses or course sequences, implementing new websites, renaming faculty and student roles, are all examples of this kind of change. In the

123

iFoundry example used throughout the manual, physical and environmental changes were not in the budget, so the team relied heavily on creating memorable names, new titles, new community elements, and so forth to push the culture in a different direction. Although physical-environmental and institutional-virtual artifacts can be used separately, working them both together is a powerful combination for effective change.

> **E-Box: Making an Israeli college more entrepreneurial.** Israel is the most entrepreneurial nation per capita on the planet, and a new president of a college of engineering in Israel came from the hi-tech sector. When he assumed the presidency of the college, one of his first acts was to eliminate individual faculty offices and create more of a hi-tech campus that mimicked those of Israel's many startups. This change helped the faculty to work together more like an entrepreneurial team at the same time it prepared students for the work environment they would face after leaving school.
>
> **Lesson.** Even modest changes in physical environment can be powerful signals for important change.

Educational Elements

We recommend starting with cultural-emotional elements in large part because it is easier to create compelling innovations in those areas. After filling in cultural-emotional elements on the canvas, we turn to educational elements themselves. Here, we work in ease-of-negotiation order, starting with those things where agreement can come quickly, holding off on the things that people want to fight about until later.

Working from the canvas template, we list the items in ease-of-negotiation order as follows:

1. Architectural elements or shifts
2. Unleashing elements
3. Elective elements

4. Supportive, extra-curricular, and community elements
5. Pedagogical enhancements
6. Intrinsic motivation enhancements
7. Required disciplinary elements

Let's look at each of these briefly.

Architectural elements or shifts

We start the educational elements with the architectural features of the program. Is this a traditional degree where core requirements and sequences of disciplinary course prerequisites lead at the end to a degree, or is the architecture somehow more modular or isolated like a series of certificate programs or coding school? In other words, how are educational experiences organized, and how do they fit together into a coherent set of structures that lead to some kind of credit, certification, or degree? What kinds of options and add-ons are available to customize degrees to individual preferences?

> **E-Box: Making a Bologna sandwich.** At a South American university, the dean of engineering faced a problem. The traditional engineering curriculum had many, many hours of required courses forced by the Ministry of Education and the national engineering societies. In moving to a curriculum aligned with the 3 year+2 year (bachelors + masters) curriculum of the European Bologna convention (European Commission, n.d.), the dean decided to open up some space for innovation by offering (1) traditional engineering degrees (Civil, Electrical, Mechanical, etc.) on the 3+2 year track that fulfilled all the many requirements of the engineering societies, and (2) some non-traditional tracks (bio-engineering & other less regulated degrees) that side-stepped the micro-management of the engineering societies.
>
> **Lesson.** Architecture can shake things up, especially if it is used creatively to offer things currently not permitted in the curriculum.

If you view your program as a traditional degree where core requirements, particular sequences of disciplinary courses, and electives, minors and other course categories, there isn't much to contemplate, but architectural innovation can be a particularly powerful way to shake things up (see the "Making a Bologna sandwich" Ebox).

Simple measures, such as introducing student choice into what were rigid disciplinary requirements ("Pick 6 of 9 of these courses") or by offering different pathways helps manage the co-contrary between *rigor* && *choice* in a way that is acceptable to faculty and more motivating to students.

Additionally, the widespread availability of course content in various MOOCs allows for ways of separating the provision of content from the assessment of student achievement in unprecedented ways. Moreover, the desire by students of the ability to shed student debt and earn now allows for novel possibilities of reorganizing curricula into bite-sized chunks that lead to employment sooner rather than later (deconstructing a degree as a series of certificates).

Is this a traditional degree where core requirements and sequences of disciplinary course prerequisites lead at the end to a degree, or is the architecture somehow more modular or isolated credit, certification, or degree like a series of certificate programs or a coding school? In other words, how are educational experiences organized, and how do they fit together into a coherent set of structures that lead to credit, certification, or degree? What kinds of options and add-ons are available to customize the educational experience to individual preferences?

Unleashing elements

A salient feature of WNE and the field manual is the notion of student unleashing. Does the educational program call on students to define and overcome a significant educational challenge or series of challenges as part of the educational process itself? Many programs are very good at defining what students should know, but not very good at unleashing students to the possibilities in their lives.

In chapter 3, we reviewed WNE's five-fold model of unleashing, the five pillars of unleashing as joy, trust, courage, openness connectedness-collaboration, and community. When Olin College had the opportunity to start from a blank slate, one way that these elements were incorporated into the curriculum was to have *design* throughout the program starting in freshman year with Design Nature, continuing in the second year with User-Oriented Collaborative Design (UOCD), and finishing in the fourth year with a yearlong Senior Capstone in Engineering (SCOPE). These courses combine constrained and open-ended work, individual and teamwork, design as an activity to help others and to solve problems, in ways that are both joyful and rigorous.

E-Box: Project-based learning as nightmare. It's easy to pay lip service to an X-learning methodology and harder to get the unleashing. On a trip not long ago, one of us was asked to visit a project-based learning class in a major South American university. Upon entering the room, three students were in front of the class ready to present the status of their project. As soon as the students opened their mouths the course professor interrupted every other sentence, correcting them along the way. Yes, feedback is important, but the students appeared humiliated, leaving the experience memorable, but not in an unleashing kind of way.

Lesson. X-learning methods can be unleashing, but don't forget to create an environment that both *challenges* && *supports* students, one where positive emotion and trust are key parts of the experience.

While design is an excellent basis for unleashing, there are others. Student thesis work challenges students to formulate and perform an investigation from start to finish. Project-based learning, cooperative learning, and experiential learning (X-learning) can be incorporated into traditional courses to manage the co-contrary between *theory* && *practice*. One caution in approaching these X-learning methods is to think of them as some automatic, step-by-step cookbook approach and to forget the joy, the courage, and the openness of the five pillars. The culture and mood of the setting is as important as the methodology.

Electives as an unleashing element

Direct ways to embed student unleashing are important, but there are other ways to *challenge* && *support* students more generally by adding to their sense of self-efficacy and motivation. In chapter 2, we briefly discussed motivation using the triple, (1) autonomy, (2) purpose/connectedness, and (3) mastery. There are a number of ways to help along these dimensions.

E-Box: Students welcoming students. An associate dean of a South American university was inspired by the examples of student community at Olin and Illinois, and he sought to create student community at his university. At first, he turned to faculty members to help, but they were busy with normal teaching and research duties, so he turned to students directly. He recruited project-ready students who had extracurricular experience in student clubs (mainly Brasil Junior *https://brasiljunior.org. br*). In short order a small group of 4 students put together an innovative program to welcome students to engineering (*Engenharia Recebe*) using passports and visa stamps and a variety of activities to connect students to the student community.

Lesson. Students are an oft-ignored source of power and light if only we treat them as resourceful, creative, and whole, and if only we involve them in important, meaningful roles in the life of the educational institution.

Students already exercise autonomy in their choice of electives, and in professional education with its rigid disciplinary pathways, additional electives in the curriculum can be a breath of motivational fresh air.

Yet sometimes students have trouble articulating their goals or purposes in allocating their electives. Put somewhat differently, they have trouble managing the co-contrary between *freedom* && *structure* in their academic decisions. Access to good academic counseling can be helpful, but we've both seen how allowing students to *give names* to small clusters of related electives (Goldberg, 2008) can help them define their purpose and choose electives in a more thoughtful manner that aligns with their career and life goals.

Supportive community & extra-curricular activities

This tension between *freedom* && *structure* can also play out in students study habits, social lives, and outside activities. Many schools now provide fairly extensive student success centers to offer counseling and tutoring services that go beyond traditional faculty office hours. Student-life operations play an important role in promoting and sustaining a healthy ecosystem of extracurricular activities.

Arrival on campus can be aided by the intentional shaping of explicit community. Referring to chapter 2, the design of Olin's small residential campus, the *partners year*, the explicit inclusion of student *passionate pursuits* on student transcripts, an integrated honor code, and copious opportunity and attention to student feedback were key elements in the design of an effective and supportive culture (WNE, Ch. 1).

In the Illinois case, the large number of faculty and students, and the small size of the iFoundry pilot demanded explicit attention to creating a sense of community through the *iCommunity*, *iTeams*, and other artifacts of the iFoundry Freshman Experience (iFX) (WNE, Ch. 2, iFoundry, 2009).

Required Disciplinary Elements

To this point we've talked about cultural-emotional elements and non-disciplinary elements in order to have fruitful conversations and large amounts of agreement without the usual arm-wrestling that accompanies discussions of required courses in discipline. At this point in the process, it is important to step back and look at all the things that are now on the table that were previously not even part of the conversation. This is already an achievement, and many of the things on the earlier portions of the canvas are things that students and other stakeholders value.

Having said this, the canvas is likely to be something of an overstuffed blintz with lots of aspiration and not much room for innovation. Moreover, this is the moment when faculty territoriality is waiting to leap in and protect "my" or "our" courses. At this point, the right thing to do is not to start suggesting specific course removal from the list of required courses. More productively, something useful to do is to suggest whether there aren't *alternative structures* for required course selection. If the list of requirements is long, is there some way to shorten the list, allowing room for electives or innovative new courses? If the list is inflexible, is there some way to allow students greater choice from a list of required courses that give them some autonomy and more control over what is after all the student's degree? If the list has a small number of specialties or options, are there ways to offer both traditional and new pathways, each with different sets of requirements? If the list has only faculty-specified pathways through the degree, is it possible to offer student-designed pathways through required courses that give students flexibility to design a package of disciplinary requirements that align with their career goals, subject to faculty oversight and approval?

In other words, at this juncture it is better to seek different options for the offering and selection of disciplinary requirements than to fight over specific course requirements themselves.

Pedagogical Enhancement

Another dimension of curriculum description is along the lines of *pedagogical enhancement*. Lecture, discussion, and lab classes have been standard for many years, but modern pedagogy (Felder & Brent, 2016) pays explicit attention to learning objectives, effective instruction technique, *active* or *experiential learning*, which may include flipped classrooms (lectures on video for homework with interactive experiences in class to master material), and cooperative learning techniques (Johnson, Johnson, & Smith, 2006) as well as enhanced methods to assess classroom effectiveness while the airplane is in flight (Angelo & Cross, 1993).

Some faculty members may already be committed to some of these methods in the courses they teach, and perhaps the desire is to get more instructors to bring more of these methods into more courses. This can entail a broad commitment to get to xx% of courses as pedagogically enhanced in some way by a 20XX, or perhaps there are some key courses (introductory courses, or senior capstones) that deserve an upgrade. At this juncture, broad aspirational targets work better than dictating that specific courses or instructors change.

Intrinsic Motivation Conversion of Required Courses

One interesting development of the iFoundry years was the development of the intrinsic motivation (IM) conversion process (Herman et al., 2017). The idea is to inject intrinsic motivational theory elements into the design of required courses. What is the societal purpose of the course? What is the disciplinary purpose of the course? What is the student's purpose for taking the course? When students are asked to reflect on this last question, they come up with a variety of answers: "I want to use it to start a business." "It's fundamental to being the great accountant I want to be" and so forth, and when students reflect on the purpose of the courses and what it means to them, they tend to be happier about the course than if they don't reflect on its meaning to them.

Also in IM conversion, students are asked to make *choices* on how they intend to demonstrate *mastery* through a process of upfront negotiation. Students can take a standard path through the course, or they can demonstrate mastery with papers, projects, alternative forms of assessment, and the like.

Moreover, with widespread availability of lecture materials for free online, it is now possible to have students attend lectures on a variety of course materials and have them evaluated at their home institution by a professor expert in the materials. Given all the online learning materials that now exist, teaching students how to fish for learning on their own may be more productive than having them catch fish, course by course.

Assessing a Canvas

Once a canvas is sketched out, it can be assessed in several ways:

1. Four spirits analysis
2. Responsiveness to student outcome profile
3. Adherence to regulatory and accreditation requirements
4. Utilization of resources.

The first assessment element is an analysis by each of the four spirits teams about the ways in which the canvas addresses the concerns of each spirit. The second element assesses responsiveness to the student outcome profile developed as well as "I-like-I-wish," data.

Next, accreditation and governmental standards are consulted, and the canvas is evaluated based on those standards. It is commonplace to elevate accreditation standards above the others, but our view is that it is important for a unit to have its own sense of what good is before it deals with accreditation and governmental requirements. We have often heard it said that "ABET won't let us do that" or "My government requires X hours of left-handed physics," when the so-called requirement is not a requirement at all. Working from stakeholder needs and druthers first, and treating accreditation and regulatory as important constraints, maximizes attention to the success of the program and satisfies regulatory and accreditation requirements, both.

Finally, the canvas addresses the feasibility and viability of the program. What new or renewed teaching resources are required for the plan? What amount of new preparation time is needed for novel or innovative elements of the plan? In what ways will the program be more or less attractive to students and employers of students? One of the unspoken advantages of the status quo is that it is often in quasi-equilibrium with these different factors, while change oftentimes requires readjustment along one or more of these dimensions. In planning for substantive change, it is important to bring necessary resources to the table, allow for faculty reinvestment in new and revised courses, and to allow sufficient time to roll out new coursework, pedagogy, and supporting programs. Understating or underestimating these real costs is a recipe for faculty unrest. We pay explicit attention to such practical matters is as an important consideration of sprint 4.

Reflexercise 6.4: What innovations can be undertaken in your program that will face little faculty resistance?

Some questions. What cultural changes can be made? What physical facilities changes can be made? What educational support activities can be invented or improved? What extracurricular activities can be added or hypertrophied? What community features can be created? Which innovations will have the greatest positive impact? What is holding back these changes? What can be done to make little bets or otherwise pilot some of these changes? What else do you notice?

Sprint 4: Negotiating Tradition and Change from Interests

In the *negotiating tradition and change sprint,* the goal is to negotiate a final proposal that manages the co-contrary between *tradition* && *change* (*status quo* && *innovation*).

To negotiate in a thoughtful way (see figure 6.6), and to manage this co-contrary, typically the design team is reduced in size to form a negotiation team with two kinds of members, those who are concerned primarily with the risks of changing and those concerned with the risks of not changing. By explicitly acknowledging this co-contrary, the negotiation is engaged in a principled manner in the hopes that the outcome will satisfy other members of the design team. There are a variety of ways to reduce the full team to a negotiating team, but the key is to teach the team members how to *negotiate from interest* rather than *negotiate from position* (Fisher, Ury & Patton, 2011).

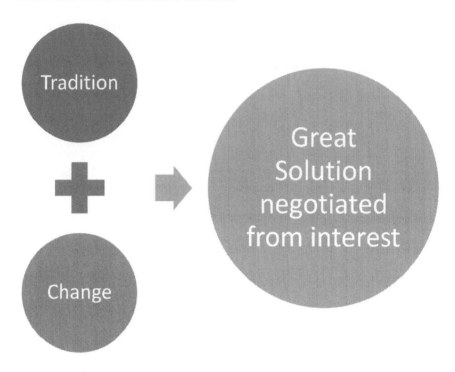

Figure 6.6: Sprint 4 negotiates a great solution from tradition and change from interest.

A Typical Negotiation from Position: A Zero-Sum Game Where You Win, Lose, or Compromise

Most of our experiences with negotiation are biased by the limited negotiation experiences we get in real life. Buying a car or dickering over the price of an object at a garage sale, the seller offers to sell high, you offer to buy low, and someone wins, loses, or you split the difference. The problems with these kinds of transactions as a model of negotiation are manifold. The object on offer is not itself subject to modification, the stakes are low, the game is zero-sum (one person's gain is the other person's loss), and the time available for negotiation is fairly limited. Any lessons learned in such situations are likely to be limited in applicability, but they form the unfortunate basis for how many of us approach more sophisticated negotiation scenarios.

Negotiation from Interest: Create More Value, Then Split the Pie

Many negotiation situations are rife with opportunities for creative give and take in which the parties to the negotiation make the pie bigger (create value) before slicing it up (claim value). Educational program design is certainly one of these situations because the object being negotiated is nuanced, complex, and inherently extensible and open to generative change, and because it is often the case that the interests of the parties along these multiple dimensions being negotiated do not necessarily need to conflict.

The natural thing to do in such situations is to shift from negotiating from position to negotiating from interest (Fisher, Ury & Patton, 2011). In this model of negotiation, attention is shifted away from the usual win-lose-compromise standard to seven elements:

1. Alternatives to negotiation
2. Interests
3. Options
4. Standards of legitimacy
5. Communication
6. Relationship
7. Commitment

Those interested in delving into the model in more detail should consult the end of chapter Dip or Dive section. Here, we briefly highlight items 2-4 starting with item 3, *options*.

The tendency for curriculum fights to devolve to a battle over required course real estate has already been ameliorated by the design of the educational model canvas. By spending canvas real estate on oft-neglected cultural-emotional and unleashing items we look to spend more time on non-controversial innovations with big impact. In this way, even if required course architecture is static, important progress can be made in short time frames with little fighting.

Another point to make is around interests. Of course, the tendency is to frame curriculum battles as a fight between faculty research factions, and one faction's gain is another's loss, but it seems to us that the framing misses the larger picture of what's being negotiated. University education is a complex affair with many stakeholders and interests, and the overemphasis on faculty and research factional interests is a fundamental flaw in the whole business. After all, if an education fails to keep up with the times, it is the student and the persons touched by the student who will be harmed by that failure; faculty and their faculty factions will get what they want out of the business, but students will pay the price of their short-sightedness. We are not saying that the curriculum should be handed over to students by any means, but we are saying that the co-contrary of *student* && *faculty* interests can be managed in a less one-sided way to the benefit of the whole enterprise.

This leads to item 4, standards of legitimacy. In principled negotiations, the parties often refer to common standards adopted in similar situations as a test of whether the parties are being fair with one another. Here, we suggest that higher education curriculum negotiations are fundamentally unfair from the standpoint of business practices in competitive marketplaces. Good businesspeople respond to customer feedback and quickly repair poor quality products or services, because in a competitive marketplace, your competitors will take away your customers if you don't. Although higher education has long been more shielded from direct competition than other businesses, the rise of online alternatives is changing the higher educational landscape in unpredictable ways. Moreover, it has become commonplace in business to survey customers and understand their likes and dislikes. This is rarely done as part of curriculum design and another fundamental flaw of the usual curriculum negotiation process is simply that outside stakeholders aren't even consulted.

Reflexercise 6.4: Reflect on a curriculum change you have been involved. These cases are often examples of "negotiation from position" in which different individuals or factions present positions and either someone wins, loses, or compromises.

Some questions. To what extent is this the case? To what extent do you find that people find common interests or make an agreement? To what extent do you believe the process resulted in winners, losers, or compromise? What else do you notice about the chosen experience?

Just-in-Time Learning & Materials

At a team level, the organization of the 4SSM is structured to shift hearts and minds in a principled manner so that the team is open to new possibilities and ways to assess success. At the level of individuals, it is also important that team members master key skills at critical moments to help promote personal shift to openness and the possibilities in change. Interestingly, at ThreeJoy, we believe that many of these same skills that are important for design team members to master should be mastered by students today as a matter of good educational and professional practice, both. Along with the team tasks outlined previously, each sprint provides just-in-time training to get CT members going on key skills and concepts.

The table below (table 4.1) shows an overview inventory of shift skills training that goes on to scaffold design team success.

Sprint	Just-in-time Shift Skills & Concepts
4 Spirits	NLQ = noticing, listening & questioning, speech acts (requests, commitments, assertions, assessments), co-contraries & the sunshine-shadow quad, Schein's 3-level culture model, intrinsic motivation, mindset, rudiments of agile & design thinking personae.
Bright Spots & Great Possibilities	Elephant, rider & path, bright spots, innovation/creativity heuristics, courage & the power of vulnerability, imposter syndrome, story reframing, the *thinking path*.
Conceptual Canvases	Business model canvas, conceptual canvas, wheel of complete communication, curriculum architecture, education pyramid
Negotiating Tradition & Innovation	Interests vs. positions models of negotiation, best-alternative-to-a negotiated agreement, (BATNA), anchoring.

Table 6.1: Each sprint has different just-in-time concepts and practices required for completion.

The first sprint is especially laden with shift skills preparation and the first sprint kickoff schedule tends to run longer than the other sprints. Thereafter, there are one or two key items that need just-in-time training to facilitate the 4SSM process.

Reflexercise 6.5: Journal or make some notes about your takeaways and action-aways from this chapter.

Some questions. Skim back over this chapter. What parts resonated with you? What parts did you disagree or have some concern with? What terms were used in ways that seemed familiar or unfamiliar? What experiences in your life connected to what you read? What questions arose in your mind for further reflection and consideration? What inspired you to action?

Dip or Dive

1. **Agile development.** *Dip.* Read the exquisitely short text *Scrum: A breathtakingly brief and agile introduction* by Hills and Johnson (2012). What do you notice about agile and scrum compared to other action frameworks you are familiar with? *Dive.* Study Atlassian's *Agile Coach* (2021) website (*https://www.atlassian.com/agile*) and do the same comparison.

2. **Design thinking.** *Dip.* Read about the process of creating personae at the website *How to create personas for design thinking* (Innovation Thinking, n.d.): *https://www.innovationtraining.org/create-personas-design-thinking/.* Create at persona for a student of the future at your educational program. *Dive.* Take a design thinking course on Coursera or some other MOOC system.

3. **Business models canvases.** *Dip.* Watch Alexander Osterwalder explain the basic notions of a business model canvas from their text *Business Model Generation* (Osterwalder & Pigneur, 2010) in this video:

https://www.youtube.com/watch?v=RpFiL-1TVLw (Darus, 2015). After a quick dip, apply the canvas to the business model of your current educational institution. *Dive.* Acquire the text mentioned above and repeat in more depth.

4. **Negotiation from interests vs. positions.** *Dip.* Watch William Ury talk about the interests behind positions in negotiation from the text *Getting to Yes* (Fisher, Ury & Patton, 2011) in the short video *https://www.youtube.com/watch?v=vdA2wecb4k0* (Ury, 2016). Consider the role of interests and positions in a negotiation with which you are familiar. *Dive.* Read the text and revisit the question.

Reflectionaire

1. **Ego-driven arm wrestling.** In what ways have you observed committee or other policy discussions devolve to ego-driven arm wrestling? What do you notice about the conclusions reached in such discussions? In what ways have you seen ego-driven arm wrestling sidestepped or overcome in productive ways? What else do you notice in thinking about this issue?

2. **Committee work.** What is the most productive thing you've ever seen a committee do? What is the least productive committee you've ever been on? What committees have you served on that adequately serve an important function? What kinds of problems do committees work well on? What kinds of problems do committees get stuck on? What else do you notice about committee work?

3. **4SSM *gedanken* experiment.** Run a thought experiment of applying the four sprints and spirits method (4SSM) at your school. How would it be received? What would be the sticking points? Would the process bog down in ego-based arm wrestling? Who do you know at your school that knows something about the component parts: design thinking, agile development, coaching, negotiation, and facilitation skills? Who would make good 4SSM design team members? Who would resist the process like the plague? What else do you notice as you reflect on this possibility?

4. **4SSM deconstruction in slow motion.** The full process requires a significant amount of commitment and resources to be done quickly. In what ways could you deconstruct the process and do it piece by piece over an extended period? What in-house resources would you have to aid this process? What external resources would you need? What else do you notice about this possibility?

5. **DIY (do it yourself).** Upon reading this chapter, what concept, experience, innovation, or other object caught your attention that deserves further reflection? What is it about this object of your attention that interests you? What have you experienced in connection with this object? What other questions need to be asked and reflected upon in connection with your DIY reflectionaire? Reflect on those questions. What else do you notice as you reflect upon this object?

Little Bets

1. **Four Spirits LB.** Take the decomposition of the four spirits and plan short unit learning activities around one of them: culture, motivation, design thinking, and co-contraries.

2. **Great possibilities LB.** Consider schools or programs that you consider to be innovative. Plan a series of unit learning activities in connection with those programs. Visit their websites and read their written materials and articles. Visit the programs in-person or on Zoom. Invite innovators from those programs to campus.

3. **Educational business model canvas little bet.** Read *Business Model Generation* (Osterwalder & Pigneur, 2010) with a group of colleagues and stakeholders. Use the book to analyze your school's current business model. What ideas do you get for business model change from reading the text? What else do you notice?

4. **Negotiation from interests.** Attend an online or in-person class on negotiation from the Harvard Law School Program on Negotiation (or comparable program). *https://www.pon.harvard. edu/executive-education/.* After attending the program, what parts

of your personal or professional life can be improved by following the principles learned?

5. **DIY little bet.** What little bet needs to be made as a result of your thinking about the material in the book so far? What is easiest to try or do with existing means? What is the smallest possible experiment that will get you some interesting results? What do you notice from doing this experiment? What else needs to be done?

7

A Process, Place & People for Affective Implementation

From Big Bang to Affective Implementation

The last chapter surveyed the rationale for and the elements of the four sprints and spirits to systematically lead a design team to a reasonable 4SSM plan for curriculum or programmatic change, but the four sprints and spirits are not the end of the story. Following the execution of the four sprints, the plan as negotiated in the final sprint is ready to be implemented, and it is important to be as thoughtful and caring in implementation as during the change design process. Unfortunately, the usual approach to implementation is to use traditional faculty leadership and committee structures to take the curriculum or programmatic plan as designed and implement it in one fell swoop, what we call *big bang implementation*. Doing so is often problematic on several grounds.

First, substantive curriculum or programmatic change can be like eating at a beautiful brunch buffet. Our aspirational eyes can be bigger than our implementation stomachs, and it may not be practical to implement everything in the plan straightaway. Yes, the third and fourth sprints try to take practicality into account, but as the rubber meets the road of a real implementation various impracticalities will come to light that were not anticipated in the four sprints process itself. Sometimes these impracticalities can be accommodated with a phased implementation plan, and more will be said about this in a moment, but it is important here (1) to acknowledge that practical difficulties in implementation are likely to arise and (2) to develop a principled approach toward those difficulties in a way that preserves both the innovation and spirit of the 4SSM process.

Second, it is common to assume that faculty leadership and committee structures can be handed the plan and are up to the implementation challenge, and with normal curriculum++ efforts this may be the case, but a substantive change process provides a greater number and variety of implementation and change challenges, and it is our recommendation that a special *space* and *team* be created for 4SSM plan implementation.

Finally, that a substantive curriculum change process like 4SSM often tees up an abundance—a surfeit—of possible changes is itself a symptom of a larger problem. Curriculum change is usually contemplated over a long cycle (five or ten years or even longer); in between times there may be coursewise innovation, but programmatic or curriculum change itself is usually performed infrequently. Since the quality revolution of the 50s, 60s, and 70s, industrial and service organizations around the world have embraced the idea of *continual improvement* or *kaizen* in which processes that can be improved are improved on an ongoing basis. If universities were to adopt *kaizen* practices toward their curricula, the long lists of implementation items generated by a substantive periodic change process like 4SSM would likely diminish in size and change wouldn't seem so daunting during period curriculum updates.

Reflexercise 7.1: Consider one or two examples of curriculum change implementation in your experience.

Reflections: How extensive were the changes (curriculum++ or substantive change)? How would you characterize the implementation process? How would you characterize the ultimate outcome? To what extent was the implementation planned, thoughtful and fair? To what extent was it ad hoc? To what extent did the work effort land unevenly or unfairly?

Given these three items, this chapter considers a different approach to implementation called *affective implementation*. The use of the term "affective" is intentional, and although we believe affective implementation is *effective*, we are speaking here of continuing the mood or emotion (the *affect*) of the four sprints into the implementation process. The big surprise of WNE was how important culture and emotion were to both the Olin and iFoundry experiences. Likewise, one of the surprising elements of 4SSM is the high quality of conversations that it engenders, and these conversations often lead to a better departmental mood, which if continued can lead to a fundamentally different departmental culture.

Thus, the unifying principle of a 4SSM-following implementation process is to pay special attention to—to prioritize—items that affect the mood and culture of the department. The idea that preserving the mood of the change process during implementation is paramount is unconventional, but it is consistent with the learning of WNE, and it is our belief that getting and keeping the mood right can make a change in student and faculty engagement in ways that will tend to overshadow even larger changes in rational curricular content.

Once the affect of the 4SSM process is acknowledged as worth preserving, it reinforces that idea of having a special place and team for implementation and continuing innovation. We will explore the creation of an *implementation team* and an *incubator* or what we call a *respectful structured space for innovation* (RSSI) for implementation and ongoing innovation as part of the affective implementation. Creating such a team and a space allows the mood of the 4SSM process to be nurtured by those who best "got" it. It also allows ideas that are not quite ripe for immediate implementation to be tested, further shaped, and refined in the hope that they can one day be fully implemented. Figure 7.1 shows the influence structure of this approach. The old culture spawns a place and a team to nurture a new mood and culture. As that mood and culture prove themselves, they increasingly back influence the old organization helping create a hybrid of the old culture and the RSSI culture.

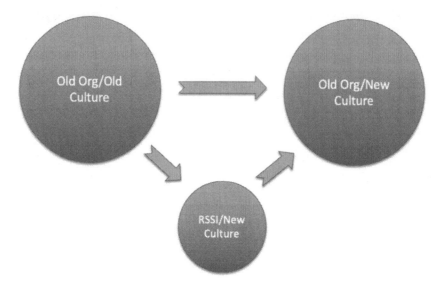

Figure 7.1: Old culture spawns RSSI which nurtures a new mood-culture that influences the old organization as the new mood-culture matures.

With a team and a space for implementation in place, it is important to examine the phased rollout of the planned changes, their communication to all stakeholders, and ways to pilot those changes not quite ready for prime time and others that arise during the ordinary course of business.

The chapter starts by examining changes in mood during the 4SSM process. It continues by considering the four elements of an affective implementation process and concludes with some tips on getting affective implemented successfully and well.

From a Battle of the Egos to Ego Distraction and Fulsome Conversation

In the last chapter, we explored eight reasons why traditional educational change fails, and a big part of the story was faculty ego. If the measure of "success" always goes back to "my curriculum vs. your curriculum" and "my research specialty vs. yours" there are few rational ways forward; the discussion is largely about power and status. One of the intentions of the 4SSM process in dividing the design team into the four spirits was to thwart this ego-driven arm wrestling by asking faculty to (1) explore unfamiliar ideas in the spirits and (2) apply them to the problem at hand. By doing this, faculty tend to converge on common solutions via different paths. In this way, faculty egos are directed away from their territorial interests and knowledge comfort zone and are directed to think intelligently about and apply the spirits to the unit and its curriculum. Of course, egos don't disappear—nor should they. Instead, they are directed to a kind of healthy competition-in-action toward the good of the unit. In various pilots of this kind of approach, we have observed the pride that goes into faculty presentations of their work product; no one necessarily wants to be the best, but no one wants to be a shlub either.

Moreover, over the course of the four sprints, repeated application of four spirits thinking by each spirit team makes each quasi-expert in something the other teams have not considered as fully. This helps keep the assessment conversations spirited, somewhat separated or distinct, and civil. Together we can think of this process as creating a kind of *ego distraction,* which does one other thing. It transforms what would been a *competitive process* ("my education was better than yours") into a more reflective and *collaborative* process ("from the perspective of culture or co-contrary or motivation or student futures, we think . . .") for putting together a sound and innovative proposal.

Another thing to notice is that the structure of the four sprints requires a lot of conversation about interesting stuff. In pilots, we have heard comments like, "I always wanted to have these kinds of conversations with my colleagues." Higher ed is a busy business, and when one becomes a professor, one imagines many intellectually stimulating conversations in the faculty lounge, but then you sign the contract letter, arrive on campus, and start jumping through tenure and promotion hoops and you realize that there doesn't seem to be much time for those imagined conversations much at all. A structured process with four key sprints divided into four spirits requires conversations by everyone about what might change, along several different dimensions, and these many conversations drive the possibility of exploration, the possibility of agreement, the possibility of caring for what others value, in a way that opens the field to the possibility of change.

The Four Sprints as a Journey to a Good Mood

The foregoing discussion does a reasonable job explaining the rational intentions and power of the 4SSM process, but a lot of the actual effect of 4SSM is affective. That is, yes, 4SSM works because it is good process and good thinking, but also because it promotes what we like to call a *journey to a good mood*. Sometimes it is easy to have a bad attitude about being a faculty member, especially in matters involving committees. Past departmental or curriculum battles weigh heavily in your thinking and feeling; returning to the same-old committee-based scheme merely picks at those old scabs and scars, causing old wounds to hurt all over again. Instead, 4SSM shifts to an unexpected process, something surprising, and it takes change team members on a journey to good mood visit by visit and sprint by sprint. Let's sketch out four steppingstones to a good mood a bit more.

Steppingstone 1: From Suspicion & Skepticism to the Surprising Possibility of Openness

During the first visit for the kickoff and start of the four spirits sprint, faculty members are suspicious, skeptical, and often in pain from their last curriculum battle. To make matters worse, they think the facilitator is a smarty pants know-it-all who is going to tell them what to do. Denials of same by the facilitator, "I don't know what you should do or what you are going to do as a group," and now they think the facilitator is merely incompetent, but during the first encounter team members are surprised by some of the activities.

None of it is the boring, directive lecturing or canned solutions they were expecting; they start to notice that they are having some interesting conversations about students, about learning, about being a faculty member. Then, they go off into their four spirits teams and start studying their assigned spirit and the ways in which it is alive and well (or not) in their program. In a word, in the first visit, there is a shift from being closed, suspicious, and skeptical to being somewhat more open to more conversation and the next step in the process.

Steppingstone 2: Sharing Observations from Spirit Learning to Caring What Others Say

During the second visit, the team members share what they learned in their four spirits teams. The mood is at first, like that of an eager freshman class, "Here's what we did. Is this okay? Did we do this right?"

Again, denying that there is a right answer, "I don't know. Did it seem important to you?" is received with funny looks. As we kick off the bright spots and great possibilities sprint, they are eager to do some "real" curriculum work "finally." But as the second visit continues, and we do more exercises and shift skills training, they start to notice that they are enjoying the conversations with their colleagues. Some say, "I always thought being a teacher would be like this, having good conversations about students, teaching, and learning." So, the mood shifts to one of confidence in conversations, caring about each other and students, with an increase in hope that something good can come out of all this work.

Steppingstone 3: Exploring Bright Spots & Great Possibilities

As the process moves along, there is palpable excitement about the bright spots and great possibilities. Many of the ideas developed by the different groups are similar to one another, giving people confidence that maybe some of the things they are important might make it to the final proposal. They are also starting to get the hang of the four spirits, and the terms "co-contrary," "culture," "motivation," and "personae," are tossed around frequently with understanding and importance. There is still some impatience with wanting to put it all together, and by the time we do the trial conceptual canvas run of the third sprint, the teams are chomping at the bit with eager anticipation to finally lay out a curriculum.

Steppingstone 4: Integrating and Negotiating the Pieces

As the sprints move along, there is a seriousness about the conceptual canvases, and some are lying in wait to fight the good fight to defend their research specialty and faction against attacks by their "opponents," but the canvas urges the teams to lay out cultural and emotional elements prior to those elements that people relish fighting about, and the fourth sprint urges agreement on emotional and curricular elements before tackling zero-sum factional debates. There are serious conversations about resources, there are serious conversations about elements that may not meet accreditation criteria, and there are concerns about changes "to my course," but there are more changes on the table, everyone feels some pain of retooling, and the final solution is negotiated.

In the end, with a properly facilitated process, there is agreement and a sense of accomplishment and a coming together. There is some concern about what the changes will bring, what it will mean personally, to the group, and to the students, but all that is backed up by the power of conversation and listening that has gone on over the various activities in the four sprints. The steps described here are not trivial, and they require more work than is expended in many curriculum or educational change processes, but the processes are direct, the conversations are purposeful, and it can avoid many of the pitfalls of other less intentional procedures.

Four Keys to the Design of Affective Implementation

Given all the conversation, given all the great curriculum work product, given the journey to a good mood, and given the raised hopes for something innovative and special, 4SSM deserves an implementation process that brings forward the rational and emotional-cultural work product of the process, both.

Shown pictorially in figure 7.2, we see the two products of the 4SSM process, the 4SSM plan and the so-called "journey to a good mood." Both are valuable, but we tend to emphasize the 4SSM plan above the good mood that was generated by the process. Doing so is a mistake, and both need to be carried over into the implementation process.

Figure 7.2 There are two outputs from 4SSM: the plan and the good mood.

Implementation & Innovation Team and the Incubator

Affective implementation starts with these two outputs—the 4SSM plan and the journey to the good mood, and carries them over to an affective implementation process by creating both a *team* and a *place* in which the mood and plan can be planted, watered, nurtured, and allowed to grow. Here we call the team the *implementation and innovation team* (IIT), and in iFoundry parlance we called such a space an *incubator* or we can use the more generic term *respectful structured space for innovation* (RSSI). Whatever names you choose, a first step toward successful implementation is to gather some of the key individuals who "got" the vibe of the mood-culture created during 4SSM and have them continue the spirit of the 4SSM-generated mood-culture as a unit in a kind of incubator or RSSI.

Figure 7.3 Schematic of an affective implementation process and its 4 elements.

Cultural Conversations, Communications and Celebration

Once the team and incubator are up and going, the next step is to continue to have conversations about the culture-mood learning of the 4SSM process and to actively communicate the 4SSM-generated plan to faculty, students, alumni, employers, and others. After an engaging 4SSM process, it is natural to breathe a sigh of relief following and return to normal work patterns, but it is important to continue to have cultural-mood conversations that continue the 4SSM learning, and to communicate the plan and the status of the implementation process systematically to others.

E-Box: Studios, studios, everywhere. What we would now call an RSSI was started in 2017 to experiment with studio learning at an Australian technical university. Building on 5 shifts training and early versions of 4SSM training, the team worked to build a more student-caring and design-oriented engineering education through *studio learning*. Starting with one studio subject reaching 14 sections totaling 120 students (2018), the team continued with summer studios in 2019 (140 students) and 2020 (160 students), but the growth in interest in studios led to schools within the faculty *pulling* studio learning into the mainstream school year. As of 2021, 82 studios in five schools have helped educate over 2500 students. Concomitant with the growth of studios is a marked increase in student engagement and a sense that the faculty cares for its students.

Lesson. By maintaining an RSSI outside of the main culture, but connected to it, successful educational experiments can be increasingly embraced by students and faculty members alike.

This is especially the case because curricula or educational programs are examples of the interesting artifacts whose reality or existence (ontology) is dependent on others believing in them. Searle (1999) has called these kinds of artifacts (objects like money, chess, or marriages) *institutional artifacts,* and communicating the details of a new curriculum to stakeholders is more than a public relations exercise. In part, the very existence of a new curriculum depends on those who might take it, teach it, or use it believing it is a "thing." Thus, bringing a new curriculum into existence is, in part, a matter of communicating its elements to others and having those people believe that it exists. Of course, these days these kinds of *ontological communication* efforts are aided by websites and social-media accounts through text posts, visual representations, as well as audio and video.

The third element of affective implementation is a *phased implementation* process. Substantive change plans can quickly generate a fair amount of work for faculty. In planning the rollout of this work, it is important to assess the amount of change, the increased workload, both initially and ongoing, and to plan a rollout that allows faculty members to keep up with their many responsibilities (teaching, research, service, and outreach). New faculty are onboarded with attention to workload and large change efforts can impose work requirements like that kind of onboarding. Certainly, experienced faculty members can make quicker work of changes than newbie faculty, but they shouldn't be expected to shoulder massive workload increases without some sort of adjustment to their work schedules.

Continued Innovation in the Incubator

The last element of affective implementation is the use of the incubator for continued innovation. As phased implementation continues, some of the planned items may require additional experimentation and work before they can be scheduled for implementation. In cases such as this, the incubator is used to run those experiments, learn, and adapt, so that the items can ultimately be implemented as part of the curriculum. Also, the incubator can be used to experiment with other novel courses or programs as they are conceived.

With this understanding of the basic design of the affective implementation process, we examine key details to get it up and running.

Steps for Getting an Affective Implementation RSSI Up and Running

The foregoing discussion makes it clear that the RSSI is the locus of implementation and continued innovation, and here we offer a series of steps in roughly implementation order to help get an effective RSSI up and running.

1. **Select team members who "get it."** Select initial implementation and innovation team members with care. Using our language from chapter 6, the IIT team should be comprised of mainly GO member and possibly some still-SLOW members, but those working on getting the 4SSM plan in place should not be bogged down with NO members who continue to resist the process.

2. **Hold a culture-mood retreat.** Hold a culture-mood retreat for one-half to several full days duration to discuss and capture what was distinctive and worth preserving about the shift in emotion and culture of the 4SSM process. The four spirits framing—culture, motivation, personae, and co-contraries—is a good place to start, and focusing on ways in which the culture-mood of the department was shifted during the 4SSM process helps understand the changes that are emerging.

3. **Create sticky name & create initial web/social media presence.** Give the RSSI a distinctive and sticky name (Heath & Heath, 2007), and create an initial web page or social media page to communicate RSSI results to stakeholders and the outside world. As was mentioned briefly earlier in the chapter, this effort is not just public relations. Communicating a new curriculum and helping others believe in it is part and parcel of the new curriculum's existence as a "thing."

4. **Make first-cut list of RSSI values.** Create a tentative list of RSSI values. The culture-mood retreat is particularly helpful for doing this. In four sprints & spirits fashion, we recognize that values often come in co-contrary pairs ("We value *rigor* and *care*"). Expressing values in co-contrary pairs is a good way of expressing the nuanced complexity of a well-designed educational program.

5. **Establish a regular RSSI meeting schedule.** Create an initial, regular schedule for RSSI meetings regardless of the current implementation needs. Doing so makes attention to the culture and mood components of the 4SSM output a priority. It also helps create community and strong relationships among those who attend.

6. **Inject beauty-love-spirit in RSSI rituals.** Begin and end RSSI meetings with distinctive rituals and get in the habit of following them. When we facilitate 4SSM or similar processes we often use Qi Gong exercises, poetry, and other non-traditional elements to start and finish our workshops. Incorporating such elements creates distinctive rituals that help identify the RSSI as special and headed in a different direction.

7. **Continue RSSI professional development around the five shifts, four sprints, and four spirits.** The 4SSM process embeds a certain amount of systematic professional development and learning around the five shifts and four sprints and spirits, but given the urgency of the project, the training is quick and cursory at best. As the affective implementation process continues, we urge teams to continue a certain amount of professional development around the five shifts, four sprints, and four spirits as well as other topics. This can be done through book reading, video viewing, podcast listening, or other means in ways that spurs conversation about what the RSSI is and is becoming.

8. **Implement the 4SSM plan in a phased manner.** Faithfully implement the 4SSM derived plan in a phased manner with attention to faculty and student workload. As was stated earlier, a big bang implementation will upset the equilibrium of the current curriculum in which faculty have balanced their workload between teaching, research, and service (and a home life) in a practical manner. Realistically assessing the work that a new curriculum requires and rolling it out in a manner that allows faculty and students to adapt will keep a department healthy and well.

9. **Take impractical plan items and establish little bets (pilots) toward practice.** In phasing in the new curriculum, certain items will simply

prove to be impractical because they require too much faculty time or other resources or because the methods assumed to be used are unknown and untested. A good portion of the reason for establishing an RSSI is to allow those good ideas to have a home for continued pilots and experimentation. Maybe there are ways to cut the resources required, or to use students in place of faculty, or to streamline or otherwise innovate to make the impractical practical. These ongoing experiments and pilots are a good source for next generation programmatic improvements.

10. **Create incentives and community within the RSSI to spur continued educational innovation.** Sometimes small streams of funding can spur educational innovation and the RSSI is a good locus for offering that funding and for housing those projects. A not insignificant benefit of having a place for innovation is the opportunity to form community among those who are predisposed to or involved in continued programmatic innovation.

Not every change effort will require all ten of these steps, but the basic idea of taking implementation seriously by having a team, a space, a communications process, and a phased plan will help ensure that the 4SSM plan lives up to its full potential.

Reflexercise 7.2: Journal or make some notes about your takeaways and action-aways from this chapter.

Some questions. Skim back over this chapter. What parts resonated with you? What parts did you disagree or have some concern with? What terms were used in ways that seemed familiar or unfamiliar? What experiences in your life connected to what you read? What questions arose in your mind for further reflection and consideration? What inspired you to action?

Dip or Dive

1. **Incubators or RSSIs.** A good practical introduction to incubator construction is in the *iFoundry Handbook* (iFoundry, 2009). *Dip.* Scan the handbook and take notes on key points that interest you in your own change initiative. *Dive.* Read the handbook carefully and take the iFoundry example of creating a board game, with roles, moves, rules, and objectives as an example for your effort, adapting and innovating as necessary to bend the example to your situation.

2. **Injected beauty for educational transformation.** Kate Goodman's (2015) dissertation suggests that beauty can be an important component of transformational educational experiences. *Dip.* Scan the workshop PDF file given at an Australian university (Goldberg, 2017) that discusses the injection of beauty. Think of ways to create transformational experiences in this way. *Dive.* Read Goodman's (2015) dissertation and repeat.

3. **DIY (do it yourself).** Upon further reflection on this chapter, what topic or topics is calling out for further reading and study? *Dip.* Do a quick web search and find a resource (post, video, podcast) that addresses this topic and give it a quick dip. *Dive.* If the topic is worthy of further study, go deeper into an original source recommended in the dipping resource (book, full-length video, or long interview). What is it about this topic that interests you? What other thoughts come to mind as you study this topic? What questions are raised or need to be asked in connection with your dip or dive. What else do you notice as you reflect upon this subject?

Reflectionaire

1. **Culture and mood as a 4SSM "output."** One unusual characteristic of this chapter is its advocacy for thinking of culture and mood as an output of the 4SSM process. What was/is your first reaction to thinking of culture and mood as an important output of the process? What assumptions of university life make it hard to think

of affect, mood, culture, or other non-tangible characteristic as an "output?" In what ways would university life be different if we were all more sensitive to culture and mood as important variables in an educational setting? What else do you notice as you consider these questions?

2. **RSSI and committees.** Reflect on the similarities and differences between RSSIs and committees. What are the similarities and differences in composition, governance, expectations, and outcomes? What else do you notice?

3. **Affective implementation *gedanken* experiment.** Run a thought experiment of applying affective implementation at your school. How would it be received? What would be the sticking points? In what ways would the progress bog down? Who would make good implementation and innovation teams members? Who would resist it like the plague? What else do you notice as you reflect on this possibility?

4. **RSSI before 4SSM.** The book has assumed that the change effort begins from a cold start, but what if you had more time to make change? Consider the possibility of creating the incubator or RSSI first. What would you call it? How would you message change ahead of starting a formal change process? In what ways would you use social media to soften up the culture? Who inside and outside your unit would you bring into the RSSI to help be most effective? What else do you notice as you consider this possibility?

5. **DIY (do it yourself).** Upon reading this chapter, what concept, experience, innovation, or other object caught your attention that deserves further reflection? What is it about this object of your attention that interests you? What have you experienced in connection with this object? What other questions need to be asked and reflected upon in connection with your DIY reflectionaire? Reflect on those questions. What else do you notice as you reflect upon this object?

Little Bets

1. **RSSI precursor LB.** Start a small reading group, elective course, club, or other informal structure to start talking about change with like-minded people. What simple activities would you undertake to get people interested in your efforts?

2. **DIY little bet.** What little bet needs to be made as a result of your thinking about the material in the book so far? What is easiest to try or do with existing means? What is the smallest possible experiment that will get you some interesting results? What do you notice from doing this experiment? What else needs to be done?

8

Taking the Manual to the Field

That was quick. We promised a short, practical field manual to get to the heart of (1) what and (2) how to change higher education (3) in a world increasingly impacted by the rise of the digital gaggle—by the rise of artificial intelligence, machine learning, robotics, and apps. Our story boils down to three short, numbered lists and an implementation process with heart: the five shifts, the four sprints, the four spirits, and affective implementation. Here we conclude by taking stock of each of these items and by offering three different paths for sequencing or using the field manual and taking it to the field.

> **Reflexercise 8.1:** Reflect on the following labels discussed in the book: digital gaggle, 5 shifts, 4 sprints, 4 spirits, affective implementation.
>
> **Some questions.** As you think back over these labels for concepts, processes, and structures, what sticks in your mind as important? Which seems less important to your situation? In what ways do these labels connect with what you already know or believe? In what ways do they challenge things you thought you knew or believed? What else do you notice as you reflect on these labels and what they stand for?

Taking Stock of the Five Shifts

Our story of the five shifts started by countering the idea that practice is the mere application of theory to particular situations—what we have called theory privilege—by arguing that (1) practice is a complex set of reflection- and conversation-in-action practices (2) that include heart and hand as well as head, (3) that treat language as something as active and generative as it is passive and descriptive, (4) that value experimentation and not knowing as much as planning and knowing, (5) that value managing paradox and co-contrary as much as solving things once and for all. In short, our starting point was to challenge the very theory- and expertise-based operating system of the university that largely treats the creation and transmission of well-vetted theory as the be- and end-all of higher education. In so doing, we shined a light on the epistemology of practice, at least in a practical way, and pointed to five key areas where the current imbalance can be addressed efficiently and effectively.

Along the way, we noticed that the five shifts were skills and mindsets where humans have a natural advantage over the digital gaggle, yet higher education persists in emphasizing exactly that material where computers have or will shortly have an advantage over human beings. Narrow, well-codified, theoretical knowledge in all fields is ripe for the continued march of the digital gaggle. By contrast, each of the five shifts connects to human *intentionality*—the characteristic of human thinking to be strongly interconnected and about other things, and today's AI is as clueless vis a vis intentionality as it was in the 80s when John Searle argued against strong AI largely on intentionality grounds (Cole, 2020). AI and machine learning do some increasingly sophisticated things, but these algorithms do not think "about" anything in any meaningful way. In this way, when humans converse and reflect about other things in multiple seemingly unrelated contexts, when they bring heart and body feelings to language, when they use speech as action as much as description, and when they navigate co-contrary and paradox with aplomb, they are doing things that computers will not for the foreseeable future. Of course, no one is arguing for the abandonment of human learning of narrow theoretical knowledge, but it does seem prudent to take key steps to AI-proof our education by spending marginally more time on exactly those skills and mindsets where humans excel over the gaggle.

Moreover, as we started to suggest in WNE (Chapter 7), and continued in chapter 4 of the field manual, the five shifts are exactly the kinds of coaching skills and mindsets that teachers need to be more effective (and affective) as human mentors and coaches in the classroom. We need faculty members to learn these things, use them and then teach them to students; the five shifts are important practices for all the constituencies of higher education.

We also suggested that the five shifts are crucial to the incubation and facilitation of human connection and insight. Our connection of the five shifts to human connection continues the call of WNE to unleash students, in part, through an emphasis on connectedness, collaboration, and community. In the field manual, we emphasized the importance of the five shifts for enhancing interpersonal empathy, understanding, and connection.

Finally, we defined insight as a kind of useful change in how we perceive or interpret incoming data, suggesting that the five shifts were especially helpful toward facilitating insight in others. Interestingly, one lesson here was that as we learn to facilitate insight in others, we learn to facilitate insight in ourselves—we learn to reflect more effectively—and the teaching sequence for facilitating insight in others both builds empathy for others, helps others, and then ultimately helps ourselves in the insight department.

Taking Stock of the Four Sprints and Spirits

Our times are ripe for educational change, and as we just argued, the five shifts are part of *what* needs to change as well as *how* that change can come about at the level of the individual instructor and student; however, change at the department, college (faculty), and university level is a team sport, and the four sprints and spirits were developed to help make curriculum or programmatic change move quickly with widespread embrace across a broad swath of faculty and other stakeholders. In this sense, 4SSM continues the call in WNE (Chapter 9) for change management and leadership processes to be carried to higher education, and in that book, we shared some of our experiences with those techniques.

In the field manual, starting from eight reasons why higher educational change typically fails, we describe a relatively quick process (weeks and months, not years) that proceeds stage by stage (sprint by sprint), encouraging faculty and other stakeholders stand back from their existing program to get more perspective about what needs to change. In one way, 4SSM mirrors (and uses) the five shifts by requiring change team members to go to the balcony and *notice* their program from a distance. The systematic progression of the four sprints, from the four spirits sprint to the bright spots and great possibilities sprint, the educational canvases sprint, and negotiation of tradition and change sprint, gradually and progressively helps faculty soften their preconceptions and look at the curriculum through multiple lenses to come to a principled and more mutually shared view of what needs to change and what needs to be kept. 4SSM is no magic wand to overcome ego or factional interests, but it does offer faculty something else to think about that can lead to greater shared interest in the change.

The strong conversational and reflective nature of 4SSM, the emphasis on co-contraries, culture, student futures, and motivation, the critical preference inputs of important stakeholders, and the negotiation of a solution that manages the co-contrary between *stability* && *change* all work together to make 4SSM outputs more innovative and yet more respectful of the status quo than many change processes. In a sense, 4SSM lives up to the ideal of the university as a place where what is done is the result of good discussions, reasoned debate, and principled negotiation. In another sense, one of the results of 4SSM is what we called a "journey to a good mood," and in that sense 4SSM deserves a special process for the implementation of 4SSM-derived educational plans.

In this way, we have come to think of 4SSM as a process toward *change that is embraced* or simply *embraced change*. Leaders will sometimes come to us and ask for ways to promote *buy-in* of this change initiative or that, but oftentimes the leader has created some initiative without much input. The 4SSM process starts, not with some notion of what needs to be done. Rather, it starts from the characteristics of a program that will be more motivating, improve student futures, manage the co-contraries and paradoxes of education in a more balanced way in ways that align with or nudge the extent culture. This starting point helps create common ground for subsequent agreement around the change initiative that emerges from deep and extensive conversations through the four sprints. In this way, there is little to react against; there is little to resist, and whatever does arise is the product of the group's interactions. Indeed, individuals, including titular leaders, can make important individual contributions, but all contributions are vetted and discussed against competing notions as well as bright spots and possibilities uncovered elsewhere. In this way, we see the process as largely collaborative, non-coercive, in ways that permit the changes to be embraced strongly across the unit.

Taking Stock of Affective Implementation

That 4SSM can lead to a plan embraced by many stakeholders is, by itself, quite a feat, and it is easy to forget that the plan is only one of two main products of all the conversation, reflection, and work of the process. What we have called a "journey to a good mood" is also quite an accomplishment, but because it is emotional and cultural in nature, it is often overlooked in favor of the more tangible and rational change plan. In chapter 7 we defined what we called affective implementation to carry over both the rational change plan and the culture-mood of the change effort.

Affective implementation takes care to (1) select the implementation and innovation team with care, (2) communicate and celebrate the culture and mood of the 4SSM effort as well as the plan, (3) implement the plan in a phased manner with concern for faculty and student workload, and (4) continue to work on those elements of the plan that turn out to be currently impractical. Given the need for culture and mood preservation of the 4SSM process, it is particularly important to create a space (physical or virtual) for the spirit of the effort to be transplanted, nurtured, and cultivated. In the field manual we have used the terms "incubator" and "respectful structured space for innovation" or RSSI for this space, but whatever term you use it is important to have a place where the culture-mood transplant can live on and continue to grow. If this is not done, a process like 4SSM turns into a kind of *faculty entertainment*—a good time had by all, but with little or no lasting effect on the culture and mood of the department. If an RSSI is created with an appropriate implementation and innovation team, the culture established during the 4SSM process can survive and can continue to work on transforming the mainline culture of the department as time goes on.

To the Field Take 1: 4SSM + Five Shifts JIT → Affective RSSI

Given these perspectives on the elements of the field manual, we now invite readers to take the manual to the field. One way to do so is to follow the advice of chapters 6 and 7 in the order presented. The four sprints and spirits or 4SSM was developed for clients trying to get the maximum change in the least amount of time, with little or no cultural preparation or pre-work. Depending upon the circumstances, this may or may not be advisable, but 4SSM as presented herein is structured to get the most change and the most cultural shift possible in the least amount of time.

Moving fast like this requires strict adherence to process, a change team willing to work in a concentrated way over a relatively short time, strong departmental/unit leadership, and skillful, attentive facilitation. By integrating just-in-time five shifts training together with the four sprints and spirits and following up with affective implementation in a substantial incubator or RSSI, substantive change can be achieved quickly and well.

The speed of this approach can also be part of its problem. If the group is working to a deadline, hiccups in one or more of the sprints can upset the process and delay completion. The need to achieve agreement quickly may result in modest changes of the existing curriculum, what we have called curriculum++. Of course, with affective implementation, the best unimplemented ideas of the 4SSM process get moved to the RSSI for continued innovation and these can come forward as full-fledged curriculum changes subsequently when they have proven to be effective and practical.

To the Field Take 2: Affective RSSI + Five Shifts → 4SSM

Another approach to using the concepts in the field manual in the world is to change the order somewhat. Recall that the changes at Olin started with faculty and students working together in the partners year and the changes at iFoundry started with faculty and students working together in the iFoundry incubator. In both cases, we may view these examples as starting with an RSSI and then moving toward a curriculum change process thereafter.

To use the field manual in this way, you would (1) create an RSSI and (2) train its members in the five shifts. Thereafter, the RSSI works to send messages about the kind of culture and curriculum that you're moving toward, take on appropriate pilot projects, and ultimately move toward more substantive curriculum change. In this reordering of the process, you create the RSSI first, train the student and faculty members in the five shifts, and at appropriate moments in time use pieces of the 4SSM to design elements of a changed program.

In the Olin case, the faculty and student partners worked to create a new curriculum from scratch. In the iFoundry case, students and faculty worked together to create a rejuvenated first-year experience across the curriculum at Illinois. Despite the difference in magnitude of the accomplishment, the creation of something with the right mood or spirit was similar in both cases.

Although the cadence of this second approach is slower than the sequence described in the field manual, it has the advantage of being robust to faculty resistance, bumpy facilitation, and uneven leadership. Taking some time before committing a curriculum to paper gives people breathing room and reflection time that allow them to reach agreement in ways that are harder to derail. Also, by starting the RSSI first and training in the five shifts, the effort explicitly recognizes the need for culture and mood change ahead of curriculum change in ways that soften the resistance to change.

To the Field Take 3: Target Culture, Mood and Experience, Not Curriculum

Usually when people express interest in educational change, the primary concern of the change leaders and team is content, curriculum, or pedagogy, and the aim of the effort, tacitly or explicitly, is to come up with a plan to change one or more of those items. Although this approach is conventional enough, our final take on using the material of the field manual suggests rising above convention and directing the field manual elements—the five shifts, 4SSM method, and affective implementation— toward shifting student experience and institutional culture and mood themselves.

This suggestion takes us back to the beginning of the field manual where we summarized the key lesson of WNE as being that culture and emotion of a program are the primary change variables that enhance student experience, learning, and unleashing, and here we end on that same point. All too often, educational change efforts assume that there exists some rational manipulation of the content, curriculum, or pedagogy of a program that will result in a better student experience and learning. As a result, the change effort ends up spending all its time shuffling course, content, and pedagogy boxes with the same deep assumptions that informed (misinformed?) the previous curriculum adjustment, assumptions such as the primacy of material coverage, the centrality of faculty expertise, and the passivity and incompetence of students. It should come as no surprise that the result of such a "change process" is often a modestly shuffled plan with little or no change in student experience or learning.

Thus, our thought in Take 3 is to use either the Take-1 or Take-2 sequencing of the material in the field manual (either sequencing will do), and apply one of them to improving the culture, mood, and student experience directly. For example, a department wanting to "revise its curriculum" might declare its existing course map, electives, required courses, options, and so forth to be fixed and given. Thereafter the unit would use the processes of the field manual with the goal of enhancing the culture, mood, and student experience of the program within the existing structure, course boxes, and requirements of the program. The five shifts training would remain the same, the four spirits teams and discussion would remain the same, the four sprints would remain the same, and affective implementation would remain the same, but the focus on changing the culture, mood, and student experience within existing constraints would have several positive effects.

First, the approach eliminates much of the drama and political tussle between departmental factions. With no turf to fight over, everyone can work toward improving student experience and unleashing within current requirements and structure.

Second, this approach should direct innovative energy toward changing the things that matter most. For example, faculty are sometimes reluctant to bring students into the discussion, because they believe students can't be trusted to make help make big changes. Because no "big changes" are on the table (or so faculty believe), faculty should more easily embrace allowing student voices into the conversation, with all the attendant benefits of doing so.

Finally, this approach generates the same "journey to a good mood" that we called out in chapter 7, but with the goal of Take 3 to change the culture and mood as well as student experience, the change team could more easily recognize that the journey to a good mood of the 4SSM process is itself a working preview of the kind of mood that is possible in a department that focuses on student learning, experience, and unleashing. By not looking for a new "plan," the mood generated during the process will be treated with the respect that it deserves, and more care will be taken to transplant the mood of the 4SSM process into the changed culture and mood of the unit.

Of course, as chapter 7 points out, this transplantation is not automatic, and we stand by the need for an RSSI, or incubator populated by team members who "get" the journey to a good mood as the place and community to cultivate and propagate this nascent culture and mood. By eliminating or reducing emphasis on the content, curriculum, and/or pedagogy plan, Take 3 puts the emphasis on the things that matter most to student experience, learning, and unleashing. Of course, there is nothing to prevent widely supported modifications to content, curriculum, or pedagogical structure and requirements from being changed, but these become a pleasant side effect—a nice to have, not a need to have—of a process that focuses on student experience, learning, and unleashing, not the main event.

To the Field We Go

Whichever *Take* you take, the five shifts, the four sprints and spirits, and affective RSSI and implementation should help your change initiative accomplish more, in less time, in ways that help bring a faculty and other stakeholders together as a group. We invite feedback on the ways in which these techniques worked or not for you, and we are especially interested in hearing stories of the kinds of changes that the manual helped you accomplish. We wish you a peaceful and productive change journey as you go to the field.

> **Reflexercise 8.2:** Journal or make some notes about your takeaways and action-aways from this field manual.
>
> **Some questions.** Skim back over the whole field manual. What parts resonated with you? What parts did you disagree or have some concern with? What terms were used in ways that seemed familiar or unfamiliar? What experiences in your life connected to what you read? What questions arose in your mind for further reflection and consideration? What inspired you to action? What little bets are important first, second, and third? What else do you notice as you reflect on these matters?

References

Adams, M. (2015). *Change your questions change your life:10 powerful tools for life and work* [3rd ed.]. San Francisco, CA: Berrett-Koehler Publishers. *https://www.amazon.com/Change-Your-Questions-Life-Leadership/ dp/162656633X*

Angelo, T. A. & Cross, K. P. (1993). *Classroom assessment techniques: A handbook for college teachers* (2nd ed.). San Francisco, CA: Jossey-Bass.

APB Speakers (2018, August 7). *James Clear's Atomic habits: How to get 1% better every day.* YouTube. *https://www.youtube.com/ watch?v=U_nzqnXWvSo*

Aristotle (1992). Nicomachean ethics. In R. McKeon (Ed.), *Introduction to Aristotle* (pp. 316-584). New York: The Modern Library. (Original work published circa 350 BCE)

Argyris, C. (1990). *Overcoming organizational defenses: Facilitating organizational learning.* Upper Saddle River, NJ: Pearson Education.

Argyris, C. & Schön, D. A. (1974). *Theory in practice: Increasing professional effectiveness.* New York: Wiley.

Argyris, C. & Schön, D. A. (1978). *Organization learning: A theory of action perspective.* Reading, MA: Addison-Wesley.

Atlassian (2021). *Agile coach. https://www.atlassian.com/agile*

Berra, Y. (1999). *The Yogi book.* New York: Workman Publishing Company.

Berrett Koehler. (2008, April 2008). *Change your questions, change your life by Marilee G. Adams* [Video]. *https://www.youtube.com/watch?v=Ahy4WfnA5AM*

Big Beacon (n.d.). *A whole new engineer. http://wholenewengineer.org.*

Big Think (2012, April 23). *Daniel Goleman introduces emotional intelligence* [Video]. Big Think. *https://www.youtube.com/watch?v=Y7m9eNoB3NU*

Brothers, C. (2005). *Language and the pursuit of happiness: A new foundation for designing your life, your relationships & your results.* Naples, FL: New Possibilities Press.

Brown, B. (2010). *The gifts of imperfection.* Center City, MN: Hazelden Publishing.

Brown, B. (2010, October 6). *The power of vulnerability* [Video]. TED Conferences. *https://www.ted.com/talks/brene_brown_the_power_of_vulnerability?language=en*

Caillet, A. (2008). The thinking path. In C. Wahl, C. Scriber, & B. Bloomfield (Eds.), *On becoming a coach: A holistic approach to coaching excellence* (pp. 149-166). New York: Palgrave Macmillan.

Cham, J. (2017). *Brain-on-a-stick* [comic]. PhD Comics. *http://phdcomics.com/comics.php?f=1126.*

Clear, J. (2018). *Atomic habits: Tiny changes, remarkable results.* New York: Avery.

Cole, D. (2020, Winter). "The Chinese room argument" in E. N. Zalta (ed.), *The Stanford encyclopedia of philosophy. https://plato.stanford.edu/cgi-bin/encyclopedia/archinfo.cgi?entry=chinese-room*

Coursera (n.d.). *Coursera: Learn without limits. http://coursera.org*

Darus, D. (2015, November 18). *Ostwalder explaining the business model canvas in 6 minutes.* YouTube. *https://www.youtube.com/watch?v=RpFiL-1TVLw*

Deci, E. & Ryan, R. M. (2018). *Self-determination theory: Basic psychological needs in motivation, development, and wellness.* New York: Guilford Press.

Dobrovolny, J. (1996). *A history of the Department of General Engineering.* Urbana, IL: University of Illinois. *https://ws.engr.illinois.edu/sitemanager/getfile.asp?id=2356*

Duhigg, C. (2012). *The power of habit: Why we do what we do in life and business.* New York: Random House.

Duhigg, C. (2013). *The power of habit: Charles Duhigg at TEDx Teachers College* [Video]. YouTube. *https://www.youtube.com/watch?v=OMbsGBlpP30*

Dweck, C. (2006). *Mindset: The new psychology of success.* New York: Random House.

Dweck, C. (2014, October 9). *Developing a growth mindset with Carol Dweck.* YouTube. *https://www.youtube.com/watch?v=hiiEeMN7vbQ*

edX (n.d.). *edX: Start learning from the world's best institutions. http://edx.org*

Elbow, P. (1983). *Writing with power: Techniques for mastering the writing process.* New York: Oxford University Press.

Elbow, P. (1987). *Embracing contraries: Explorations in learning and teaching.* New York: Oxford University Press.

Emerson, B. & Lewis, K. (2019). *Navigating polarities: Using both/and thinking to lead transformation.* Washington, DC: Paradoxical Press.

European Commission (n.d.). *The Bologna process and the European higher education area. https://ec.europa.eu/education/policies/higher-education/bologna-process-and-european-higher-education-area_en*

Feigenbaum, E. A. & Feldman, J. (Eds.) (1963). *Computers and thought.* New York: McGraw-Hill.

Felder, R. M. & Brent, R. (2016). *Teaching and learning STEM: A practical guide.* San Francisco, CA: Jossey-Bass.

Fisher, R., Ury, W. & Patton, B. (2011). *Getting to yes: Negotiating agreement without giving in* (3rd rev. ed.). New York: Penguin.

Flores, F. (1981). *Management and communication in the office of the future* [Unpublished doctoral dissertation]. University of California, Berkeley.

Gardner, H. (1993). *Multiple intelligence: The theory in practice.* New York: Basic Books.

Gardner, H. (2009, November 7). *Howard Gardner of the multiple intelligences theory* [Video]. YouTube. *https://www.youtube.com/watch?v=l2QtSbP4FRg*

Georgia TechTalks (2009). *David Rock: Your brain at work.* [Video]. YouTube. *https://www.youtube.com/watch?v=XeJSXfXep4M*

Goldberg, D. E (1989), *Genetic algorithms in search, optimization, and machine learning,* Reading, MA: Addison-Wesley.

Goldberg, D. E. (2002). *The design of innovation: Lessons from and for competent genetic algorithms.* Boston: Kluwer Academic Publishers.

Goldberg. D. E. (2008). *Human artifacts, processes, & interactions (HAPI) themes. Unpublished manuscript. https://www.dropbox.com/s/kc7gncvgc4tecvh/HAPI-10-12-08.doc?dl=0*

Goldberg, D. E. (2015, February 10). *No, slow & go: 3 speeds of transformation.* LinkedIn. *https://www.linkedin.com/pulse/slow-go-3-speeds-transformation-dave-goldberg/*

Goldberg, D. E. (2016, May 9) Effectual entrepreneurship: An interview with Saras D. Sarasvathy [Audio podcast episode]. In *Big Beacon Radio.* VoiceAmerica. *https://www.voiceamerica.com/episode/92149/effectual-entrepreneurship-an-interview-with-saras-d-sarasvathy*

Goldberg, D. E. (2016, November 6). Little bets and breakthrough ideas: An interview with Peter Sims [Audio podcast episode]. In *Big Beacon Radio.* VoiceAmerica. *https://www.voiceamerica.com/episode/95621/little-bets-and-breakthrough-ideas-an-interview-with-peter-sims*

Goldberg, D. E. (2016, December 19). Language and the pursuit of happiness: An interview with Chalmers Brothers. In *Big Beacon Radio*. VoiceAmerica. *https://www.voiceamerica.com/episode/96378/language-and-the-pursuit-of-happiness-an-interview-with-chalmers-brothers*

Goldberg, D. E. (2017). *Transformative experience: How to facilitate beauty-in-action in utilitarian subjects.* Unpublished workshop deck presented at UTS. ThreeJoy Associates, Inc. *https://www.dropbox.com/s/dx3o6aqo8dcaw5y/Transformation-Workshop-UTS.pptx.pdf?dl=0*

Goldberg, D. E. (2018, November 13). *AI's big blind spot* [PDF of PowerPoint slides presented at GIST AI Forum]. *https://www.dropbox.com/s/v55316htvcbvg3i/GIST-deg-2018-final.pptx.pdf?dl=0*

Goldberg, D. E. (2018, December 10). *5+1 shifts to the deep missing basics.* Linkedin. *https://www.linkedin.com/pulse/51-shifts-deep-missing-basics-david-e-goldberg/*

Goldberg, D. E. (2019a). *The 4 sprints method.* Unpublished proposal. Douglas, MI: ThreeJoy Associates. (related post at *https://threejoy.com/services/4sm/*)

Goldberg, D. E. (2019b, March 2019). *Shift #2: Conversation in action—Taking soft skills seriously.* Linkedin. *https://www.linkedin.com/pulse/shift-2-conversation-in-action-taking-soft-skills-david-e-goldberg/*

Goldberg, D. E. & Somerville, M. (2014). *A whole new engineer: The coming revolution in engineering education.* Douglas, MI: ThreeJoy Associates, Inc.

Goleman, D. (1985). *Vital lies, simple truths: The psychology of self-deception.* New York: Simon & Schuster Paperbacks.

Goleman, D. (1995). *Emotional intelligence: Why it can matter more than IQ.* New York: Bantam Books.

Goodman, K. (2015). *The transformative experience in engineering education.* [Unpublished doctoral dissertation]. University of Colorado. *https://www.proquest.com/openview/7a2969504f1b18f2a3a2 1a9fc060aee9/1?pq-origsite=gscholar&cbl=18750*

Harvard Law School: Program on Negotiation (n.d.). *Executive education.* *https://www.pon.harvard.edu/executive-education/*

Heath, C. & Heath, D. (2007). *Made to stick: Why some ideas survive and others die.* New York: Random House.

Heath, C. & Heath, D. (2010). *Switch: How to change when change is hard.* New York: Broadway.

Heifetz, R. & Linsky, R. (2002, June). *A survival guide for leaders. Harvard Business Review. https://hbr.org/2002/06/a-survival-guide-for-leaders*

Herman, G. L., Goldberg, D. E., Trenshaw, K. F., Somerville, M. & Stolk, J. (2017). The intrinsic-motivation course design method. *International Journal of Engineering Education, 33*(2), 558-574.

Hills, C. & Johnson, H. L. (2012). *Scrum: A breathtakingly brief and agile introduction.* San Mateo, CA: Dymaxicon.

Holland, J. H. (1975). *Adaptation in natural and artificial systems.* Ann Arbor, MI: University of Michigan Press.

iFoundry (2009). *iCommunity & iTeam handbook: academic year 2009-2010.* iFoundry. *https://www.slideshare.net/ifoundry/icommunity-handbook*

Innovation Training. (n.d.). *How to create personas for design thinking.* *https://www.innovationtraining.org/create-personas-design-thinking/*

Johnson, B. (2014). *Managing polarities: Identifying and managing unsolvable problems.* Amherst, MA: HRD Press.

Johnson, B. (2020). *And: Making a difference by leveraging polarity, paradox or dilemma* (Vol. 1). Amherst, MA: HRD Press.

Johnson, D. W., Johnson, R. T., & Smith, K. A. (2006). *Active learning: Cooperation in the college classroom.* Edina, MN: Interaction Book Company.

Kaplan, J. (2020, September 1). *John Searle's Chinese Room* [Video]. YouTube. *https://www.youtube.com/watch?v=tBE06SdgzwM*

Klein, G. (1998). *Sources of power: How people make decisions.* Cambridge, MA: MIT Press.

Kotter, J. (1996). *Leading change.* Boston, MA: Harvard Business School Press.

Kotter, J. & Rathgeber, H. (2016). *That's not how we do it here!: A story about how organizations rise, fall—and can rise again.* New York: Portfolio.

Laing, R. D. (1970). *Knots.* New York: Pantheon Books.

Minerva University (n.d.) *Minerva University: Higher education with purpose. http://minerva.edu.*

Mosior, B. (2021, February 24). Donald A. Schön at Iowa State University [Video & talk transcript]. Hired Thought. *https://hiredthought.com/2021/02/24/donald-a-schon-at-iowa-state-university-talk-transcript/*

Osterwalder, A. & Pigneur, Y. (2010). *Business model generation: A handbook for visionaries, game changers, and challengers.* New York: John Wiley & Sons.

Outlier (n.d.). *Outlier: Online college reimagined. http://outlier.org.*

Ozenc, K. & Hagan, M. (2019). *Rituals for work: 50 ways to create engagement, shared purpose, and a culture that can adapt to change.* Hoboken, NJ: Wiley.

Paul Quinn College (2021). *Work program. https://pqc-edu.squarespace.com/work-program*

Pink, D. (2009). *Drive: The surprising truth of what motivates us.* New York: Riverhead.

Polarity Partnerships (n.d.). *The polarity approach to continuity and transformation (PACT)™. https://assessmypolarities.com.*

Polarity Partnerships (2020, December 20). *Thriving USA video short by Barry Johnson* [Video]. Vimeo. *https://vimeo.com/489320397.*

Quote Investigator (2018). *In theory there Is no difference between theory and practice, while in practice there is. https://quoteinvestigator. com/2018/04/14/theory/*

Rogers, C. R. (1969). *Freedom to learn.* Columbus, OH: Charles E. Merrill Publishing Company.

Rock, D. (2009). *You brain at work: Strategies for overcoming distraction, regaining focus, and working smarter all day long.* New York: HarperCollins.

RSA Animate (2010, April 1). *Drive: The surprising truth of what motivates us* [Video]. YouTube. *https://www.youtube.com/watch?v=u6XAPnuFjJc*

Sarasvathy, S. D. (2008). *Effectuation: Elements of entrepreneurial expertise.* Northampton, MA: Edward Elgar Publishing.

Schein, E. H. (2009). *The corporate culture survival guide.* New York: John Wiley and Sons.

Schein, E. H. (2013). *Humble inquiry: The gentle art of asking instead of telling.* Oakland, CA: Berrett-Koehler Publishers, Inc.

Schön, D. (1983). *The reflective practitioner: How professionals think in action.* New York: Basic Books.

Searle, J. (1999). *Mind, language, and society.* New York: Basic Books.

Seligman, M.E.P. (1991). *Learned optimism.* New York: Houghton Mifflin

Sims, P. (2011). *Little bets: How breakthrough ideas emerge from small discoveries.* New York: Free Press.

Smith, R. A. (2019). *Rage inside the machine: The prejudice of algorithms, and how to stop the internet making bigots of us all.* London: Bloomsbury.

Sniady, K. (2012, September 3). *Levels of listening 1* [Video]. YouTube. *https://www.youtube.com/watch?v=x3DnB_nhGFM*

Stanford Alumni (2019, November 1). *Take charge: Design your workplace culture with Kursat Ozenc.* YouTube. *https://www.youtube.com/watch?v=dAroNh-P1UM*

Toulmin, S. (2001). *Return to reason.* Cambridge, MA: Harvard University Press.

Turkle, S. (2012, February). *Connected, but alone?* [Video]. TED Conferences. *https://www.ted.com/talks/sherry_turkle_connected_but_alone?language=en*

Udacity (n.d.). *Udacity: Radical talent transformation is here.* *http://udacity.com*

Ury. W. (2016, May 28). *Interests behind negotiating positions* [Video]. YouTube. *https://www.youtube.com/watch?v=vdA2wecb4k0*

Wang, R. W. (2021, July 25). *Yinyang (Yin-yang).* The Internet Encyclopedia of Philosophy. *https://iep.utm.edu/yinyang/.*

Whitworth, K., Kimsey-House, K., Kimsey-House, H., & Sandahl, P. (2009). *Co-active coaching: New skills for coaching people toward success in work and life* (2nd ed.). Boston, MA: Davies-Black.

About Dave and Mark

DAVID E. GOLDBERG (Dave) is an artificial intelligence pioneer, engineer, entrepreneur, educator, and leadership coach (Georgetown). Author of the widely cited Genetic Algorithms in Search, Optimization, and Machine Learning (Addison-Wesley, 1989) and co-founder of ShareThis.com, in 2007 he started the Illinois Foundry for Innovation in Engineering Education (iFoundry). In 2010, he resigned his tenure and professorship at the University of Illinois to work full time for the improvement of higher education. Dave now gives motivational workshops and talks, consults with educational institutions around the globe, and coaches individual educators and academic leaders to bring about timely, effective, and wholehearted academic change.

MARK SOMERVILLE is a professor of electrical engineering and physics at Olin College, where he also serves as Provost. He joined the founding team at Olin from Vassar College in 2001. An electrical engineer and liberal arts double major, Rhodes scholar, and semiconductor researcher, Mark remains passionate about active learning and student engagement. At Olin, Mark played pivotal roles ranging from the design of curriculum to the creation of its Summer Institute and Collaboratory to the invention of Olin's distinctive reappointment and promotion process. Mark has collaborated with established institutions such as Harvard and TUDelft as well as new programs, including Fulbright University in Vietnam, the Woodrow Wilson Academy in Boston, and NMITE in the UK.

About ThreeJoy® Associates, Inc.

ThreeJoy Associates®, Inc was founded in 2010 in the wake of the success of the iFoundry initiative at the University of Illinois. Using the methods of the field manual—the five shifts, four sprints and spirits method, and affective implementation—can benefit from a variety of ThreeJoy® services to help bring about substantive change at your educational institution:

- **Inspirational speaking.** Get your institution excited about change by exposing colleagues to inspirational stories from change experiences around the globe. Get your team oriented to the emotional-cultural shifts as much as the rational content, curriculum, and pedagogical shifts of traditional change methodologies.
- **Strategic planning and effectuation consultation.** Get your change initiative underway by focusing on key goals and change processes that matter.
- **Experienced and humble facilitation.** Get help in bringing about the change that you believe is necessary with just-in-time facilitation and training in key sprints, spirits, shifts, and implementation that bring your initiative to life.
- **Deep listening and coaching (for individuals or teams).** Being listened to in a deep way can be immensely empowering. ThreeJoy® coaching will help key members of the change team or the change team itself identify and overcome difficult challenges as well as find and unravel important change opportunities.

- **Five shifts, 4SSM, and affective implementation training.** Prefer to do it yourself? ThreeJoy® personnel can train your team in the five shifts, four sprints and spirits, and/or affective implementation so you can run your own change initiative with little or no ongoing help.

For more information, write *info@threejoy.com* today.

Made in United States
Orlando, FL
09 January 2024